D0026183

High Energy Astrophysics

FRONTIERS IN PHYSICS

David Pines, Editor

Volumes of the Series published from 1961 to 1973 are not officially numbered. The parenthetical numbers shown are designed to aid librarians and bibliographers to to check the completeness of their holdings.

Titles published in this series prior to 1987 appear under either the W.A. Benjamin or the Benjamin/Cummings imprint; titles published since 1986 appear under the Addison-Wesley imprint.

(1)	N. Bloembergen	Nuclear Magnetic Relaxation: A Reprint Volume, 1961
(2)	G. F. Chew	*S*-Matrix Theory of Strong Interactions: A Lecture Note and Reprint Volume, 1961
(3)	R. P. Feynman	Quantum Electrodynamics: A Lecture Note and Reprint Volume, 1961 (8th printing, 1983).
(4)	R. P. Feynman	The Theory of Fundamental Processes: A Lecture Note Volume, 1961 (6th printing, 1980)
(5)	L. Van Hove, N. M. Hugenholtz, and L. P. Howland	Problems in Quantum Theory of Many-Particle Systems: A Lecture Note and Reprint Volume, 1961
(6)	D. Pines	The Many-Body Problem: A Lecture Note and Reprint Volume, 1961 (6th printing, 1982)
(7)	H. Frauenfelder	The Mössbauer Effect: A Review—with a Collection of Reprints, 1962
(8)	L. P. Kadanoff and G. Baym	Quantum Statistical Mechanics: Green's Function Methods in Equilibrium and Nonequilibrium Problems, 1962 (5th printing, 1981)
(9)	G. E. Pake	Paramagnetic Resonance: An Introductory Monograph, 1962 [cr. (42)—2nd edition]
(10)	P. W. Anderson	Concepts in Solids: Lectures on the Theory of Solids, 1963 (5th printing, 1982)
(11)	S. C. Frautschi	Regge Poles and *S*-Matrix Theory, 1963
(12)	R. Hofstadter	Electron Scattering and Nuclear and Nucleon Structure: A Collection of Reprints with an Introduction, 1963
(13)	A. M. Lane	Nuclear Theory: Pairing Force Correlations to Collective Motion, 1964
(14)	R. Omnès and M. Froissart	Mandelstam Theory and Regge Poles: An Introduction for Experimentalists, 1963
(15)	E. J. Squires	Complex Angular Momenta and Particle Physics: A Lecture Note and Reprint Volume, 1963
(16)	H. L. Frisch and J. L. Lebowitz	The Equilibrium Theory of Classical Fluids: A Lecture Note and Reprint Volume, 1964
(17)	M. Gell-Mann and Y. Ne'eman	The Eightfold Way: (A Review—with a Collection of Reprints), 1964
(18)	M. Jacob and G. F. Chew	Strong-Interaction Physics: A Lecture Note Volume, 1964
(19)	P. Nozières	Theory of Interacting Fermi Systems, 1964

FRONTIERS IN PHYSICS

David Pines, Editor (*continued*)

(20)	J. R. Schrieffer	Theory of Superconductivity, 1964 (revised 3rd printing, 1983)
(21)	N. Bloembergen	Nonlinear Optics: A Lecture Note and Reprint Volume, 1965 (4th printing, 1981)
(22)	R. Brout	Phase Transitions, 1965
(23)	I. M. Khalatnikov	An Introduction to the Theory of Superfluidity, 1965
(24)	P. G. deGennes	Superconductivity of Metals and Alloys, 1966
(25)	W. A. Harrison	Pseudopotentials in the Theory of Metals, 1966 (2nd printing, 1971)
(26)	V. Barger and D. Cline	Phenomenological Theories of High Energy Scattering: An Experimental Evaluation, 1967
(27)	P. Choquard	The Anharmonic Crystal, 1967
(28)	T. Loucks	Augmented Plane Wave Method: A Guide to Performing Electronic Structure Calculations—A Lecture Note and Reprint Volume, 1967
(29)	Y. Ne'eman	Algebraic Theory of Particle Physics: Hadron Dynamics in Terms of Unitary Spin Currents, 1967
(30)	S. L. Adler and R. F. Dashen	Current Algebras and Applications to Particle Physics, 1968
(31)	A. B. Migdal	Nuclear Theory: The Quasiparticle Method, 1968
(32)	J. J. J. Kokkedee	The Quark Model, 1969
(33)	A. B. Migdal and V. Krainov	Approximation Methods in Quantum Mechanics, 1969
(34)	R. Z. Sagdeev and A. A. Galeev	Nonlinear Plasma Theory, 1969
(35)	J. Schwinger	Quantum Kinematics and Dynamics, 1970
(36)	R. P. Feynman	Statistical Mechanics: A Set of Lectures, 1972 (7th printing, 1982)
(37)	R. P. Feynman	Photo-Hadron Interactions, 1972
(38)	E. R. Caianiello	Combinatorics and Renormalization in Quantum Field Theory, 1973
(39)	G. B. Field, H. Arp, and J. N. Bahcall	The Redshift Controversy, 1973 (2nd printing, 1976)
(40)	D. Horn and F. Zachariasen	Hadron Physics at Very High Energies, 1973
(41)	S. Ichimaru	Basic Principles of Plasma Physics: A Statistical Approach, 1973 (2nd printing, with revisions, 1980)
(42)	G. E. Pake and T. L. Estle	The Physical Principles of Electron Paramagnetic Resonance, 2nd Edition, completely revised, enlarged, and reset, 1973 [cf. (9)—1st edition]

FRONTIERS IN PHYSICS

David Pines, Editor (*continued*)

Volumes published from 1974 onward are being numbered as an integral part of the bibliography:

Number		
43	R. C. Davidson	Theory of Nonneutral Plasmas, 1974
44	S. Doniach and E. H. Sondheimer	Green's Functions for Solid State Physicists, 1974 (2nd printing, 1978)
45	P. H. Frampton	Dual Resonance Models, 1974
46	S. K. Ma	Modern Theory of Critical Phenomena, 1976 (5th printing, 1982)
47	D. Forster	Hydrodynamic Fluctuations, Broken Symmetry, and Correlation Functions, 1975 (3rd printing, 1983)
48	A. B. Migdal	Qualitative Methods in Quantum Theory, 1977
49	S. W. Lovesey	Condensed Matter Physics: Dynamic Correlations, 1980
50	L. D. Faddeev and A. A. Slavnov	Gauge Fields: Introduction to Quantum Theory, 1980
51	P. Ramond	Field Theory: A Modern Primer, 1981 (4th printing, 1983)
52	R. A. Broglia and A. Winther	Heavy Ion Reactions: Lecture Notes Vol. I: Elastic and Inelastic Reactions, 1981
53	R. A. Broglia and A. Winther	Heavy Ion Reactions: Lecture Notes Vol. II, *in preparation*
54	Howard Georgi	Lie Algebras in Particle Physics: From Isospin to Unified Theories, 1982
55	P. W. Anderson	Basic Notions of Condensed Matter Physics, 1983
56	Chris Quigg	Gauge Theories of the Strong, Weak, and Electromagnetic Interactions, 1983
57	S. I. Pekar	Crystal Optics and Additional Light Waves, 1983
58	S. J. Gates, M. T. Grisaru, M. Roček, and W. Siegel	Superspace *or* One Thousand and One Lessons in Supersymmetry, 1983
59	R. N. Cahn	Semi-Simple Lie Algebras and Their Representations, 1984
60	G. G. Ross	Grand Unified Theories, 1984
61	S. W. Lovesey	Condensed Matter Physics: Dynamic Correlations, 1986
62	P. H. Frampton	Gauge Field Theories, 1986
63	J. I. Katz	High Energy Astrophysics, 1986

High Energy Astrophysics

JONATHAN I. KATZ

Department of Physics and McDonnell Center for the Space Sciences
Washington University, St. Louis

Addison-Wesley Publishing Company, Inc.
The Advanced Book Program
Menlo Park, California • Reading, Massachusetts
Don Mills, Ontario • Wokingham, U.K. • Amsterdam • Sydney
Singapore • Tokyo • Madrid • Bogota • Santiago • San Juan

Sponsoring Editor: Richard W. Mixter
Production Editor: Karen Gulliver
Production Assistant: Connie J. Sorensen

Library of Congress Cataloging-in-Publication Data

Katz, Jonathan I.
High energy astrophysics.

Includes index.
1. Nuclear astrophysics. 2. Stars. I. Title.
QB464.K37 1986 523.01'9 86-13670

ISBN 0-201-11830-0

BCEDFGHIJK-MA-8987

Addison-Wesley Publishing Company, Inc.
2725 Sand Hill Road
Menlo Park, California 94025

To my teachers, who taught me what I know;

to my wife, who spurred me on;

to my children, who give me hope.

Contents

Editor's Foreword

The problem of communicating in a coherent fashion recent developments in the most exciting and active fields of physics continues to be with us. The enormous growth in the number of physicists has tended to make the familiar channels of communication considerably less effective. It has become increasingly difficult for experts in a given field to keep up with the current literature; the novice can only be confused. What is needed is both a consistent account of a field and the presentation of a definite "point of view" concerning it. Formal monographs cannot meet such a need in a rapidly developing field, while the review article seems to have fallen into disfavor. Indeed, it would seem that the people most actively engaged in developing a given field are the people least likely to write at length about it.

FRONTIERS IN PHYSICS was conceived in 1961 in an effort to improve the situation in several ways. Leading physicists frequently give a series of lectures, a graduate seminar, or a graduate course in their special fields of interest. Such lectures serve to summarize the present status of a rapidly developing field and may well constitute the only coherent account available at the time. Often, notes on lectures exist (prepared by the lecturer himself, by graduate students, or by postdoctoral fellows) and are distributed in mimeographed form on a limited basis. One of the principal purposes of the FRONTIERS IN PHYSICS Series is to make such notes available to a wider audience of physicists.

It should be emphasized that lecture notes are necessarily rough and informal, both in style and content; and those in the series will prove no exception. This is as it should be. One point of the series is to offer new, rapid, more informal, and, it is hoped, more effective ways for physicists to teach one another. The point is lost if only elegant notes qualify.

As FRONTIERS IN PHYSICS has evolved, a third category of book, the informal text/monograph, an intermediate step between lecture notes and formal texts or monographs, has played an increasingly important role in the series. In an informal text or monograph an author has reworked his or her lecture notes to the point at which the manuscript represents a coherent summation of a newly-developed field, complete with references and problems, suitable for either classroom teaching or individual study.

During the past two decades, high energy astrophysics, the study of physical phenomena in the vicinity of compact or collapsed astronomical objects, such as neutron stars, black holes, and degenerate dwarf stars, has emerged as a distinct sub-field of physics and astronomy. In the present volume, Jonathan Katz, one of its leading practitioners, provides the interested non-specialist with a lucid introduction to the basic ingredients of theoretical astrophysics, the tools of non-equilibrium thermodynamics and hydrodynamics. He shows how these enable the theoretical astrophysicist to model the high energy phenomena encountered in the formation and evolution of neutron stars and degenerate dwarf stars, and in accreting black holes and quasars. I share his belief that "High Energy Astrophysics" can be read with profit by any reader with a working knowledge of undergraduate physics. It is a pleasure to welcome Jonathan Katz as a contributor to FRONTIERS IN PHYSICS.

David Pines

Urbana, Illinois
July 1986

Preface

This book grew out of a one quarter or one semester course I taught at UCLA and at Washington University, and I have tried to retain the informality of lecture notes. It is meant to be explanatory and expository, rather than complete or definitive. My students were advanced undergraduates or beginning graduate students; the reader should be fluent in undergraduate physics, but need know no astronomy or astrophysics.

The expression "high energy astrophysics" means different things to different people; the contents therefore reflect my interests, and the comments my opinions. My purpose is to describe the ingredients, methods, and results of modern astrophysical phenomenology and modelling. This is mostly the study of phenomena discovered in the last quarter of a century, and involving compact or collapsed objects. I specifically exclude cosmology, general relativity, and the detailed theories of the interiors of degenerate stars, but review the classical theory of stellar structure, which is the foundation of much of modern astrophysics.

Most of this book presents a few basic results, principles, and illustrations which all interested scientists should know. I hope these will be useful for some time. The remainder expands upon their implications and applications. Occasionally I try to offer a new point of view or make a speculative suggestion, but the bulk of the text rests upon firmer ground. In the last chapter I describe the understanding of some observed phenomena as it now exists.

I have cited the research literature only when it is of historical interest, or when necessary to support a specific assertion. Instead, I have provided citations to texts and to recent review articles. The reader who plans to begin research in this field will need to turn to the current literature; any attempt to survey it would soon become obsolete.

In writing a book I learned how much they are abstracted from earlier work. It is inevitable that a book on this subject draw on two excellent recent works, *Radiative Processes in Astrophysics*, by Rybicki and Lightman, and *Black Holes, White Dwarfs, and Neutron Stars*, by Shapiro and Teukolsky, which I commend to the reader. I have been more indirectly influenced by two older books, *Structure and Evolution of the Stars*, by Schwarzschild, and *Astrophysical Concepts*, by Harwit, whose scientific style was part of my education. I owe them a debt also.

This book was largely written when I was a guest of the Department of Nuclear Physics, The Weizmann Institute of Science, Rehovot, Israel. I thank them for their hospitality, Washington University for a grant of sabbatical leave, and the United States-Israel Educational Foundation for a Fulbright Lectureship. I also thank T. Piran for comments, my editors for applying just the right amount of pressure, and my wife for a careful reading of the manuscript.

אומברה אס קם
איחו קריה
ארבול פלנטה
ליברו אסקריבה

High Energy Astrophysics

Chapter 1

Stars

1.1 Generalities

This book was not meant to be about stars. But stars are the most familiar, best studied, and arguably most important objects in the astrophysicist's universe. They are therefore the building blocks of many theories of more exotic objects. More fundamentally, the study of stars is the study of the competition between gravity and pressure. Astrophysics is distinguished from nearly all of the rest of physics by the importance of gravity, so that an understanding of the principles of stellar structure is necessary in order to understand most other astronomical objects.

The study of stellar structure and evolution is an elaborate and mature subject. The underlying physical principles are mostly well-known, and have been developed in great detail. Powerful numerical methods produce quantitative results for the properties and evolution of stars. Numerous texts and a very extensive research literature document this field. I refer the reader to three standard texts; although not new they have aged very well, and it would be both pointless and presumptuous to attempt to improve on them. Chandrasekhar (1939) reviews the classical mathematical theory of stellar structure, whose beginnings are now more than a century old. Schwarzschild

(1958) presents a less mathematical description of the physical principles of stellar structure and evolution, with more attention to the observed phenomenology. This is probably the best book for a general introduction to the properties of stars and their governing physics. I recommend it (supplemented by any of the numerous recent descriptive astronomy books) as a reference for the physicist without astronomical background. Clayton (1968) is particularly concerned with processes of nucleosynthesis and thermonuclear energy generation.

There are still a number of outstanding problems in the theory of ordinary stars. Many of these arise from a single area of theoretical difficulty: the problem of quantitatively describing turbulent flows. This problem arises in the formation of stars from diffuse gas clouds, in stellar atmospheres, for rotating stars and accretion discs (which may be thought of as the limiting case of rapidly rotating stars), in interacting binary stars, in stars with surface abundance anomalies, and in stellar collapse and explosion. If turbulent flows have a material effect on the properties of a star, quantitative theory must usually be supplemented by rough approximations, and confident calculation becomes uncertain and approximate phenomenology. This is even more true of the more exotic objects which are the subject of this book.

The problems of turbulent flow appear in two distinct forms. In the first form, a turbulent flow arises in an otherwise well-understood configuration, and may even resemble the turbulent flows known to hydrodynamicists; the problem is the calculation of some property, usually an effective transport coefficient, of the flow. The most familiar example of this is turbulent convection in the solar surface layers. In the second form, the initial or boundary conditions of a flow are not known; it may not be turbulent in the hydrodynamicist's sense of eddies or nonlinear wave motion on a broad range of length scales, but quantitative calculation is still impossible. The formation of stars is an example of this kind of flow. A variety of assumptions, approximations, and models, generally of uncertain validity and unknown accuracy, are used to study turbulent flows in astrophysics.

This chapter on stars has two purposes. One is to illustrate some of those physical principles of stellar structure which are useful in understanding stars and other astrophysical objects. The other is to develop the kind of rough (often order-of-magnitude) estimates and dimensional analysis which are widely used in modelling novel astrophysical phenomena. Some of this material follows Schwarzschild (1958).

1.2 Phenomenology

Hundreds of years of observations of stars have produced an enormous body of data and revealed a wide variety of phenomena which are discussed in numerous texts and monographs and a voluminous research literature. Here we will summarize only the tiny fraction of those data essential to the astrophysicist who wishes to use stars in models of high energy astrophysical phenomena.

The luminosities and surface temperatures of stars are often described by their place on a Hertzsprung-Russell diagram, such as that shown in Figure 1.1. In this theoretician's version the abscissa is the stellar effective surface temperature T_e, defined as the temperature of a black body which radiates the same power per unit area as the actual stellar surface; the ordinate is the stellar photon luminosity in units of the Solar luminosity $L_\odot = 3.9 \times 10^{33}$ erg/sec. There are also observers' versions in which the abscissa is a "color index," a directly observable measure of the spectrum of the emitted radiation, and the ordinate may be the absolute or apparent stellar magnitude in some observable part of the spectrum. Accurate conversion between these

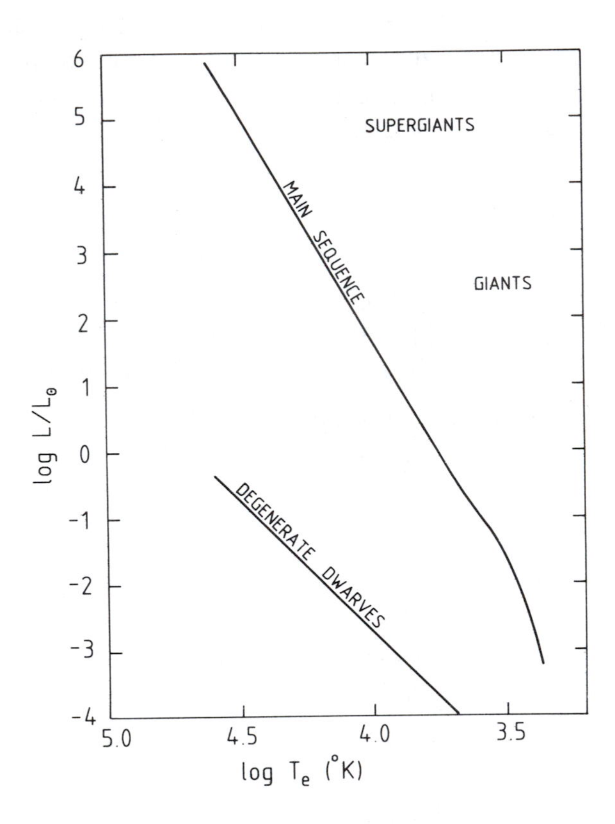

Figure 1.1. Hertzsprung-Russell diagram.

two versions requires a quantitative knowledge of the spectrum of emitted radiation, which is approximately (but not exactly) that of a black body.

Most stars are found to lie on a narrow strip called the main sequence. These stars (occasionally referred to as dwarves) produce energy by the thermonuclear transmutation of hydrogen into helium near their centers. Their positions along the main sequence are determined by their masses, which vary monotonically from about $30M_\odot$ (where the solar mass $M_\odot = 2 \times 10^{33}$ gm) at the upper left to $0.1M_\odot$ in the lower right. The Sun lies on the main sequence near its middle.

Stars found above and to the right of the main sequence are called giants and supergiants; their higher luminosities (and their names) are accounted for by large radii, ranging in extreme cases up to 10^{14} cm, about 1000 times that of the Sun. These stars have exhausted the hydrogen at their centers and produce energy by thermonuclear reactions in shells close to, but outside, their centers. Stars of nearly equal ages (such as the members of a single cluster of stars, formed nearly simultaneously) will be distributed along a narrow track in the giant and supergiant region, a track whose form reflects their complex evolutionary path. Stars of a broad range of ages, such as the totality of stars in the solar neighborhood, will mostly be found on the main sequence; those in the giant and supergiant regions will be broadly distributed rather than lying on a narrow track. There are no sharp distinctions among main sequence (dwarf) stars, giants, and supergiants, and intermediate cases are found.

Degenerate (traditionally called white) dwarves are faint, dense stars in whose interiors the electrons are Fermi-degenerate, resembling the state of an ideal metal or metallic liquid. They generally produce negligible thermonuclear energy, having converted essentially all their hydrogen (and probably also their helium) to heavier elements. Their meager luminosity is supplied by their thermal energy content, possibly augmented by the latent heat of crystallization, the gravitational energy released by the sedimentation of their heavier elements, and other minor sources. They cool steadily as these energy sources are exhausted. Degenerate dwarves move to the lower right along a track parallel to lines of constant radius as they cool. Their radii depend on their masses (roughly as their reciprocals), but because their masses are believed to span a moderate range (perhaps $0.4M_\odot$ to $1.2M_\odot$) they all lie in a strip of moderate width. These masses are less than those which these stars had when young, but the amount of mass lost is controversial and may range from a few percent of to nearly all the initial mass. It is not known whether the mass in the degenerate dwarf stage is a monotonic function of or even determined by the mass at birth; it may be random and unpredictable. Very few stars other than degenerate dwarves are found much below and to the left of the main sequence; most of these few are probably evolving rapidly into degenerate dwarves.

An extrapolation of the main sequence to the lower right leads to stars of mass too low to produce thermonuclear energy, generally called brown dwarves. These objects slowly evolve into degenerate dwarves of very low mass and lie near (but above, because of their low masses) an extrapolation of the degenerate dwarf strip. Jupiter may be regarded as an extreme case. These objects are nearly unobservable because of their low luminosities, and only a few, if any, can be identified with confidence. Their properties are uncertain because the properties of matter under brown dwarf conditions are not well known; few data are available to test the uncertain calculations.

Objects at the upper left end of the main sequence are very rare, with their rarity increasing with increasing mass and luminosity. As a consequence, extrapolation beyond masses of $50M_\odot$ is largely limited to theory.

1.3 Equations

A star may be defined as a luminous self-gravitating gas cloud. If it is also spherical, in hydrostatic equilibrium, and in thermal steady state it is described by the classical equations of stellar structure:

$$\frac{dP(r)}{dr} = -\frac{\rho(r)GM(r)}{r^2} \tag{1.3.1}$$

$$\frac{dM(r)}{dr} = 4\pi r^2 \rho(r) \tag{1.3.2}$$

$$\frac{dL(r)}{dr} = 4\pi r^2 \rho(r)\epsilon(r) \tag{1.3.3}$$

$$\frac{dT(r)}{dr} = -\frac{3\kappa(r)\rho(r)L(r)}{16\pi acT^3(r)r^2}. \tag{1.3.4}$$

Here $P(r)$ is the pressure, $M(r)$ is the mass enclosed by a sphere of radius r, $\rho(r)$ is the density, $L(r)$ is the luminosity produced within a sphere of radius r, $\epsilon(r)$ is the rate of nuclear energy release per gram, $T(r)$ is the temperature, $\kappa(r)$ is the Rosseland mean opacity (defined in **1.7.2**) in cm^2/gm, and a is the radiation constant. The first three of these equations are elementary; (1.3.4) is derived in **1.7**.

Numerous assumptions and approximations have been made: spherical symmetry, Newtonian gravity, a star in a stationary (unchanging) state, and a flow of energy by the diffusion of radiation only. Various of these assumptions may be relaxed if the equations are appropriately modified. It is frequently necessary to allow for the transport of energy by turbulent convection (most familiarly, in the outer layers of the Sun) or by conduction (in electron-degenerate matter).

These equations must be supplemented by three constitutive relations, derived from the microscopic physics of the stellar material. For any given chemical composition they take the form:

$$P = P(\rho, T) \tag{1.3.5}$$

$$\epsilon = \epsilon(\rho, T) \tag{1.3.6}$$

$$\kappa = \kappa(\rho, T). \tag{1.3.7}$$

These equations of stellar structure may be solved numerically, which is necessary to obtain quantitative results. It is illuminating, however, to make order-of-magnitude estimates. If we did not have computers available (and were unwilling to integrate these equations numerically by hand), or did not know the quantitative form of the constitutive relations, these rough estimates would be the best that we could do. Until the development of quantitative theories of thermonuclear reactions and opacity, no detailed calculation was possible. Even today, rough estimates are the basis of most qualitative understanding. In novel circumstances they are the first step toward building a quantitative model.

1.4 Estimates

1.4.1 Order of Magnitude Equations In order to make rough approximations to the differential equations (1.3.1–4) we replace them by algebraic equations in which the variables P, M, L, and T represent their mean or characteristic values in the star, the continuous variable r is replaced by the stellar radius R, and the derivative d/dr is replaced by the multiplicative factor $1/R$. In most cases this level of approximation produces useful rough results, although it is occasionally disastrous; with intelligent choice of the numerical constants it can be remarkably accurate, though usually only when a quantitative solution is available as a guide.

The equations become:

$$P = \rho \frac{GM}{R} \qquad (1.4.1)$$

$$M = \frac{4}{3}\pi R^3 \rho \qquad (1.4.2)$$

$$L = \frac{4}{3}\pi R^3 \rho \epsilon \qquad (1.4.3)$$

$$T^4 = \frac{3\kappa\rho L}{16\pi a c R}. \qquad (1.4.4)$$

We now assume the perfect nondegenerate gas constitutive relation for pressure

$$\begin{aligned} P &= P_g + P_r \\ &= \frac{\rho N_A k_B T}{\mu} + \frac{aT^4}{3}, \end{aligned} \tag{1.4.5}$$

where P_g and P_r are the gas and radiation pressures respectively, μ is the mean molecular weight (the number of atomic mass units per free particle), N_A is Avogadro's number per gram, k_B is Boltzmann's constant, and a is the radiation constant. Combination of (1.4.1), (1.4.2), and (1.4.5) (ignoring the radiation pressure term in 1.4.5, an excellent approximation for stars like the Sun) yields results for the characteristic values of ρ, P, and T:

$$\rho = \frac{3M}{4\pi R^3} \tag{1.4.6}$$

$$P = \frac{3GM^2}{4\pi R^4} \tag{1.4.7}$$

$$T = \frac{GM}{R}\frac{\mu}{N_A k_B}. \tag{1.4.8}$$

1.4.2 Application to the Sun In Table 1.1 we compare the numerical estimates for ρ, P, and T obtained by substituting the solar mass, radius, and molecular weight, to the quantitative values found for the center of the Sun in a numerical integration (Schwarzschild 1958) of the equations (1.3.1)–(1.3.7). More recent calculations (Bahcall, *et al.* 1982) produce slightly different numbers, but the difference is of no importance when we are examining the validity of order-of-magnitude estimates. We use $R = 6.95 \times 10^{10}$ cm, $M = 2 \times 10^{33}$ gm, and $\mu = 0.6$.

The estimated value of T is remarkably accurate (probably fortuitously so), while the estimates of ρ and P are low by two orders of

Table 1.1

	Estimate	Solar Center
ρ (gm/cm^3)	1.42	134
P (dyne/cm^2)	2.73×10^{15}	2.24×10^{17}
T (°K)	1.39×10^7	1.46×10^7
κ (cm^2/gm)	2.18×10^3	1.07
ϵ (erg/gm/sec)	1.95	14

magnitude (note that the estimated ρ is nothing more than the mean stellar density). This large discrepancy reflects the concentration of mass towards the center of a star, and is a consequence of the compressibility of gases and the inverse-square law of Newtonian gravity. The discrepancy also reflects a deliberate obtuseness on our part in comparing the estimated values of ρ and P to the calculated central values. Had we been more cunning we could have chosen to compare to a suitable chosen "mean" point in the numerical integration, and would have obtained truly impressive (but deceptive) agreement.

In all stars the central density greatly exceeds the mean density. In stars of similar structure this ratio is nearly constant, and the greatest use of eqs. (1.4.6–8) is as scaling relations among stars of differing mass and radius. Rough estimates and qualitative understanding may be obtained readily; numerical integrations are always possible when quantitative results are needed.

For giant and supergiant stars the ratio of central to mean density may be as much as 10^{16}. Such enormous ratios indicate a complete breakdown of the approximations (1.4.1–4); the interior structure of such stars is very different from that of stars like the Sun; it can be roughly described by simple relations, but requires an understanding of their peculiar structure. In fact, their condensed central cores and very dilute outer layers may each be separately described by equations (1.4.6–8) with reasonable accuracy; disaster strikes only when one attempts to describe both these regions together.

Equations (1.4.3) and (1.4.4) may also be used to estimate κ and ϵ given the estimates for ρ, P, and T. For the Sun we use $L = 3.9 \times 10^{33}$ erg/sec. These numerical values are also compared in Table 1.1 to quantitative values at the Solar center (Schwarzschild, 1958). The estimated value of ϵ is just the Solar (mass-weighted) mean; the actual central value is several times higher because thermonuclear reaction rates are steeply increasing functions of temperature, which peaks at the center. The estimated value of κ is far wrong; this is in part because of the hundredfold concentration of density at the center, and in part because of the concentration into a small central core of thermonuclear energy generation. Equation (1.3.4) shows that using an erroneously low estimated ρ and high R produces an erroneously large estimate for κ.

Except for temperature, our rough estimates have been very inaccurate. Approximations like those of equations (1.3.6) and (1.3.7) are still useful, particularly when only scaling laws are needed for a qualitative understanding. They can also produce semiquantitative results when some additional understanding is inserted into the equations in the form of intelligently chosen numerical coefficients. We have deliberately refrained from doing so in order to show the pitfalls as well as the utility of rough estimates; when aided by intuition and guided by experience they can do much better.

1.4.3 Minimum and Maximum Stellar Surface Temperatures The observed range of stellar surface temperatures is approximately 2500°K to 50,000°K. These limits each have simple explanations.

The continuum opacity of stellar atmospheres is largely attributable to bound-free (photoionization) and free-free (inverse bremsstrahlung) processes. For the visible and near-infrared photons carrying most of the black-body flux at low stellar temperatures the most important bound-free transition is that of the H^- ion, which has a threshold of 0.75 eV. At temperatures of a few thousand degrees matter consists largely of neutral atoms and molecules, and the small equilibrium (Saha equation) free-electron density is very sensitive to temperature, dropping precipitously with further decreases in temperature. The H^- abundance, in equilibrium with the free electrons, drops nearly as steeply. The atmosphere approaches the very transparent molecular gas familiar from the Earth's atmosphere. As a consequence of this steep drop in opacity, the photosphere (the layer in which the emitted radiation is produced) of a very cool star forms at a temperature around 2500°K, below which there is hardly enough opacity and emissivity to absorb or emit radiation. This temperature bound is insensitive to other stellar parameters, and amounts to an outer boundary condition on integrations of the stellar structure equations for cool stars.

The maximum stellar surface temperature has a different explanation. In luminous stars the radiation pressure far exceeds the gas pressure, and the luminosity is nearly the Eddington limiting luminosity L_E (**1.11**), at which the outward force of radiation pressure equals the attraction of gravity:

$$L \approx L_E \equiv \frac{4\pi cGM}{\kappa}, \tag{1.4.9}$$

where κ is the opacity. Under these conditions the opacity is predominantly electron scattering, and $\kappa = 0.34\ \mathrm{cm}^2/\mathrm{gm}$, essentially independent of other parameters. The effective (surface) temperature T_e is then approximately given by

$$T_e^4 = \frac{cGM}{\kappa\sigma_{SB}R^2}, \tag{1.4.10}$$

where σ_{SB} is the Stefan-Boltzmann constant. In order to estimate R we approximate the pressure by the radiation pressure

$$P \approx \frac{a}{3}T^4, \tag{1.4.11}$$

where T is an estimate of the central temperature. Note that here we neglect the gas pressure; in obtaining equation (1.4.8) we neglected the radiation pressure. Eliminating P and ρ from (1.4.1), (1.4.6), and (1.4.11) produces an estimate for R:

$$R^4 = \frac{9GM^2}{4\pi a T^4}. \tag{1.4.12}$$

Substituting this result in (1.4.10) gives

$$T_e^4 = \frac{T^2 c}{\kappa\sigma_{SB}}\sqrt{\frac{4}{9}\pi acG}. \tag{1.4.13}$$

Because thermonuclear reaction rates are usually very steeply increasing functions of temperature, the condition that thermonuclear energy production balances radiative losses acts as a thermostat; detailed calculation shows that $T \approx 4 \times 10^{7}$°K, nearly independent of other parameters for these very massive and luminous stars. Numerical evaluation of (1.4.13) then gives

$$T_e \approx 90{,}000^\circ K. \tag{1.4.14}$$

This numerical value is about twice as large as the results of detailed calculations, but they confirm the qualitative result of a mass-independent upper bound to T_e for hydrogen burning stars.

1.5 Virial Theorem

For stars (defined as self-gravitating spheres in hydrostatic equilibrium) it is easy to prove a virial theorem, so named because it is closely related to the virial theorem of point-mass mechanics. Begin with the equation (1.3.1) of hydrostatic equilibrium and assume it is always valid:

$$-\rho(r)\frac{GM(r)}{r^2} = \frac{dP(r)}{dr}. \tag{1.5.1}$$

Multiply each side by $4\pi r^3$, and integrate over r, integrating by parts:

$$\begin{aligned} -\int_0^R \rho(r)\frac{GM(r)}{r}4\pi r^2 dr &= \int_0^R \frac{dP(r)}{dr}4\pi r^3 dr \\ &= -\int_0^R 12\pi r^2 P(r)dr + 4\pi r^3 P(r)\Big|_0^R. \end{aligned} \tag{1.5.2}$$

The definition of the stellar radius R is that $P(R) = 0$. Hence

$$-\int_0^R \rho(r)\frac{GM(r)}{r}4\pi r^2 dr = -3\int_0^R P(r)4\pi r^2 dr. \tag{1.5.3}$$

The left hand side is the integrated gravitational binding energy of the star E_{grav}. For a gas which satisfies a relation $P \propto \rho^\gamma$ for adiabatic processes we can use the thermodynamic relation (see **1.9.1**)

$$P = (\gamma - 1)\mathcal{E}, \tag{1.5.4}$$

where $\mathcal{E}$ is the internal energy per unit volume. If we denote the integrated internal energy content of the star by E_{in} we obtain

$$E_{grav} = -3(\gamma - 1)E_{in}. \tag{1.5.5}$$

Denoting the total energy $E = E_{in} + E_{grav}$ we have

$$E = E_{in}(4 - 3\gamma) = E_{grav}\left(\frac{3\gamma - 4}{3\gamma - 3}\right) \leq 0. \tag{1.5.6}$$

The inequality comes from the requirement that a star be energetically bound. This simple relation is very useful in qualitatively understanding stellar stability and energetics.

For perfect monotonic nonrelativistic gases (including the fully ionized material which constitutes most stellar interiors) $\gamma = 5/3$; this applies even if the electrons are Fermi-degenerate. For a perfect gas of relativistic particles or photons $\gamma = 4/3$; this is a good description of gases whose pressure is largely that of radiation. Gases in which new degrees of freedom appear as the temperature is raised (for example, those undergoing dissociation, ionization, or pair production) may have still lower values of γ, approaching 1. Interatomic forces reduce γ if attractive, or increase it if repulsive (as for the nucleon-nucleon repulsion of neutron star matter).

If $\gamma = 5/3$, as is accurately the case for stars like the Sun, and more roughly so for most degenerate (white) dwarves and for neutron stars, then $E = \frac{1}{2}E_{grav} = -E_{in} < 0$. Such a star is gravitationally bound with a large net binding energy, and resists disruption. It is also stable and resists dynamical collapse, because in a smaller and denser state $|E_{grav}|$ and $|E|$ would be larger. In order to reach such a state it would have to reduce its total energy E, but on dynamical time scales energy is conserved. Energy can only be lost by slow radiative processes (including emission of neutrinos); in most cases it is stably replenished from thermonuclear sources.

A star with $\gamma > 4/3$ may be thought of as having negative specific heat, because an injection of energy increases E, which reduces $|E|$, $|E_{grav}|$ and E_{in} (see 1.5.6). Because temperature is a monotonically increasing function of E_{in} (and depends only on E_{in} for perfect nondegenerate matter) this injection of energy leads to a reduction in temperature; similarly, the radiative loss of energy from the stellar surface, if not replenished internally, leads to increasing internal temperature. The reason for this somewhat surprising behavior, described as a negative effective specific heat, is the fixed relation (1.5.5) between E_{in} and E_{grav}, which holds so long as the assumption of hydrostatic equilibrium is strictly maintained. The negative effective specific heat is also the reason thermonuclear energy release, which increases rapidly with temperature, is usually stably self-regulating.

In a degenerate star the relation between E_{in} and temperature is complicated by the pressure of a Fermi energy and the effective specific heat is positive when thermonuclear or radiative processes are considered; thermonuclear energy release is either insignificant or unstable, and radiation produces steady cooling. On dynamical time scales processes are adiabatic and the star is stable just as is a nondegenerate star. E_{in} is related to the Fermi energy which is proportional to the temperature for adiabatic processes, and the effective specific heat is again negative.

A star with $\gamma = 4/3$ has $E = 0$; the addition of 1 erg is sufficient to disrupt it entirely, and the removal of 1 erg to produce collapse. Of course, stars with γ exactly equal to 4/3 do not exist (and cannot exist, for this reason), but as γ approaches 4/3 a star becomes more and more prone to various kinds of instability. Stars with γ very close to 4/3 include very massive stars whose pressure is almost entirely derived from radiation, and degenerate dwarves near their upper mass (Chandrasekhar) limit.

A star with $\gamma < 4/3$ would have positive energy and would be exploding or collapsing. Such stars do not exist, but localized regions with $\gamma < 4/3$ do. They are found in cool stellar atmospheres (especially those of giants and supergiants) in which matter is partly ionized, and possibly in the cores of evolved stars which are hot enough for thermal pair production or dense enough for nuclei to undergo inverse β-decay. Such regions tend to destabilize a star, though the response of the entire star must be calculated to determine if it is unstable; instability is a property of an entire star in hydrostatic equilibrium, not of a subregion of it.

1.6 Time Scales

A star is characterized by a number of time scales. The shortest is the hydrodynamic time scale t_h, which is defined

$$t_h \equiv \sqrt{\frac{R^3}{GM}}. \tag{1.6.1}$$

This is approximately equal to the time required for the star to collapse if its internal pressure were suddenly set to zero. The fundamental mode of vibration has a period comparable to t_h, as does a circular Keplerian orbit skimming the stellar surface. For phenomena with time scale much longer than t_h the star may be considered to be in hydrostatic equilibrium, and eq. (1.3.1) applies. On shorter time scales the application of (1.3.1) is in general not justified. For the Sun $t_h \approx 26$ minutes.

The thermal time scale t_{th} is defined

$$t_{th} \equiv \frac{E}{L}, \tag{1.6.2}$$

where E is the total energy (gravitational plus internal) of the star, as defined in **1.5**, and L is its luminosity. This is the time which would be required for a star to substantially change its internal structure if its thermonuclear energy supply were suddenly set to zero. For phenomena with time scales longer than t_{th} the star may be considered to be in thermal equilibrium, and eq. (1.3.3) applies. The application of (1.3.3) on shorter time scales is in general not justified. For the Sun $t_{th} \approx 2 \times 10^7$ years.

The longest time scale is the thermonuclear time t_n, defined by

$$t_n \equiv \frac{M\varepsilon c^2}{L}, \tag{1.6.3}$$

where εc^2 is the energy per gram available from thermonuclear reactions of stellar material. This measures the life expectancy of a star in a state of thermal equilibrium. After a time of order t_n its fuel will be exhausted and its production of radiant energy will end; a wide variety of ultimate fates are conceivable, including cooling to invisibility, explosion, and gravitational collapse. For ordinary stellar composition $\varepsilon \approx 0.007$; about 3/4 of this is accounted for by the conversion of hydrogen to helium and about 1/4 by the conversion of helium to heavier elements. For the Sun $t_n \approx 10^{11}$ years; its actual life will be about ten times shorter because after the exhaustion of the hydrogen in a small region at the center, L will begin to increase rapidly and its remaining life will be brief. The Sun is presently near the midpoint of its life.

There is an additional time scale t_E which characterizes stars in general. In **1.4.3** we saw that there is a characteristic luminosity L_E (Eq. 1.4.9) which serves as an upper bound on stellar luminosities. Define the Eddington time t_E as the thermonuclear time t_n for a hypothetical star of luminosity L_E. Then

$$t_E \equiv \frac{\varepsilon c \kappa}{4\pi G} = \frac{2\varepsilon e^4}{3Gc^3 m_e^2 m_p \mu_e}, \tag{1.6.4}$$

where we have written the electron scattering opacity κ in terms of fundamental constants and μ_e is the mean number of nucleons per electron. For ordinary stellar composition $t_E \approx 3 \times 10^6$ years. This is an approximate lower bound on the lifespan of a star. Because it nearly four orders of magnitude shorter than the age of the universe, luminous stars have passed through many generations, manufacturing nearly all the elements heavier than helium. The luminosities of stars range over at least nine orders of magnitude, so lower luminosity stars have lifetimes very much longer than t_E, and even much longer than the present age of the universe.

A quantity analogous to the Eddington time is also an important parameter in the study of rapidly accreting masses (for example, in models of X-ray sources and quasars; Salpeter 1964). The luminosity is given by $L = \dot{M}c^2\varepsilon$. The Salpeter time is defined as the e-folding time of the mass M, if $L = L_E$:

$$t_S \equiv \frac{M}{\dot{M}} = \frac{\varepsilon c\kappa}{4\pi G}. \tag{1.6.5}$$

It is usually estimated that $\varepsilon \sim .1$, so that $t_S \sim 4 \times 10^7$ years. This is the characteristic lifetime of such a luminous accreting object.

Finally, there is a simple "light travel" time scale t_{lt} which may be defined for any object of size R:

$$t_{lt} \equiv \frac{R}{c}. \tag{1.6.6}$$

It is generally not possible for an object of size R to substantially change (by a factor of ~ 2) its emission on a time scale shorter than t_{lt}, because that is the shortest time in which signals from a single triggering event can propagate throughout the object, and hence the shortest time on which its emission can vary coherently. A small change, by a factor $1 + \delta$ with $\delta \ll 1$, can occur in a time $\sim \delta t_{lt}$. If the velocity of propagation were the sound speed (or, equivalently, a free-fall speed) rather than c, then t_{lt} would be the hydrodynamic time t_h given by (1.6.1).

The time scale t_{lt} is chiefly used in models of transient or rapidly variable objects in high energy astrophysics, such as variable quasars and active galactic nuclei, γ-ray bursts, and rapidly fluctuating X-ray sources. The observation of a substantial variation in the radiation of an object in a time t_{var} is evidence that its size R satisfies

$$R \lesssim ct_{var}. \tag{1.6.7}$$

Such an upper bound on R may then be combined with the luminosity to place a lower bound on the radiation flux and energy density within the object, and therefore to constrain models of it.

These arguments contain loopholes. It is possible to synchronize clocks connected to energy release mechanisms and distributed over a large volume so that they all simultaneously trigger a sudden release of energy (because the clocks are at rest with respect to each other there is no difficulty in defining simultaneity). A distant observer would not see the energy release to be simultaneous, but rather spread over a time t_{lt}, where R is the difference in the path lengths between him and the various clocks. However, if the clocks have appropriately chosen delays which cancel the differences in path lengths, he will see the signals of all the clocks simultaneously, violating (1.6.7). This would require a conspiracy among the clocks which is unlikely to occur except by intelligent design, and would produce a signal violating (1.6.7) only for observers in a narrow cone.

Other loopholes are more likely to occur in nature. A strong brief pulse of laser light propagating through a medium with a population inversion depopulates the excited state at the moment of its passage. Nearly all of the medium's stored energy may appear in a thin sheet of electromagnetic energy, whose thickness may be much less than R, and whose duration measured by an observer at rest may violate (1.6.7). This is a familiar phenomenon in the laser laboratory, in which nanosecond (or shorter) pulses of light may be produced by arrays of lasing medium more than a meter long.

Analogous to a thin sheet of laser light is a spherical shell of relativistic particles streaming outward from a central source (Rees 1966). If they produce radiation collimated outward (radiation produced by relativistic particles is usually directed nearly parallel to the particle velocity) the shell of particles will be accompanied by a shell of radiation. This radiation shell will propagate freely, and will eventually sweep over a distant observer, who may see a rapidly varying source of radiation whose duration violates (1.6.7). The factor by which it is violated depends on the detailed kinematics of the radiating particles. In general, (1.6.7) is inapplicable when there is bulk relativistic motion, even if only of energetic particles; conversely, its violation implies bulk relativistic motion.

1.7 Radiative Transport

1.7.1 Fundamental Equations The most important means by which energy is transported in astrophysics is by the flow of radiation from regions of high radiant energy density to those of lesser; radiation carries energy from stellar interiors to their surfaces, and from their surfaces to dark space. The complete theory of this process is unmanageably and incalculably complex and cumbersome, but a variety of approximations make it tractable and useful. Fortunately, these approximations are well justified in most (but not all) circumstances of interest, so that the theory is not only tractable but also powerful and successful. Here we will be concerned principally with the simplest limit, applicable to stellar interiors, in which matter is dense and opaque, and radiation diffuses slowly. There is another, even simpler limit, that of vacuum, through which radiation streams freely at the speed c. Between these limits there are the more complex problems of radiative transport in stellar atmospheres (by definition, the regions in which the observed photons are produced). This is a large field of research blessed with an abundance of observational data; several texts exist (for example, Mihalas 1978).

Consider in spherical coordinates the propagation of a beam of radiation, so that r measures the distance from the center of the coordinate system and ϑ is the angle between the beam and the local radius vector. In general, the radiation intensity I will depend on the point of measurement (r, θ, ϕ) (note that ϑ must be distinguished from the polar angle θ), on the polarization, and the the photon frequency ν. In most cases it is possible either to assume spherical symmetry (so that there is no dependence on θ and ϕ), or to treat the problem at different θ and ϕ locally, so that these angles enter only as parameters of the solution, like the chemical composition of the star being studied. In either case it is not necessary to consider θ and ϕ explicitly, and they will be ignored, along with any dependence of the intensity on the azimuthal angle φ of its propagation direction. Problems in which these approximations are not permissible are difficult, and generally their solution requires Monte Carlo methods (in which the paths of large numbers of test photons are followed on a computer in order to determine the mean flow of radiation). I also neglect polarization because it does not significantly affect the flow of radiative energy; it is worth calculating in some stellar atmospheres because it is sometimes observable for nonspherical stars or during eclipses (symmetry implies that the radiative flux integrated over the surface of a spherical star is unpolarized). The frequency dependence of the radiation field is important, although it will not always be written explicitly.

In travelling a small distance dl a beam loses a fraction $\kappa\rho dl$ of its intensity, where κ is the mass extinction coefficient (with dimensions of $\mathrm{cm}^2/\mathrm{gm}$), and ρ is the matter density. We consider a beam with intensity $I(r,\vartheta)$ (with dimensions $\mathrm{erg/cm^2/sec/steradian}$, where the element of solid angle refers to the direction of propagation, not to the geometry of the spherical star); the power crossing an element of area ds normal to the direction of propagation, and propagating in an element $d\Omega$ of solid angle, is $I(r,\vartheta)dsd\Omega$. In the short path dl a power $I(r,\vartheta)\kappa\rho dldsd\Omega$ is removed from the beam by matter in the right cylinder defined by ds and dl, where we have taken $d\Omega \ll ds/dl^2$. Matter also emits radiation, and the volume emissivity j is defined so that the power emitted by the volume $dlds$ into the beam solid angle $d\Omega$ is $j\rho dlds\frac{d\Omega}{4\pi}$. The units of j are erg/gm/sec and the emission is assumed isotropic, as is the case unless there is a very large magnetic field.

After travelling the distance dl the radiation field transports energy out of the cylinder with a power $I(r+dr, \vartheta+d\vartheta)dsd\Omega$, where it has been essential to note that a straight ray (we neglect refraction) changes its angle to the local radius vector as it propagates. In a steady state the energy contained in the cylinder does not change with time, so that the sum of sources and sinks is zero:

$$I(r,\vartheta)dsd\Omega - I(r,\vartheta)\kappa\rho dl dsd\Omega + j\rho dl ds\frac{d\Omega}{4\pi} - I(r+dr,\vartheta+d\vartheta)dsd\Omega = 0. \tag{1.7.1}$$

From elementary geometry

$$dr = dl\cos\vartheta \tag{1.7.2a}$$

$$d\vartheta = -dl\sin\vartheta/r. \tag{1.7.2b}$$

These equations are a complete description of the trivial problem of the propagation of a ray in vacuum, and may be combined and integrated to yield the solution

$$r = r_\circ \csc\vartheta, \tag{1.7.3}$$

where $r_\circ$ is the distance of closest approach of the ray to the center of the sphere. If the polar axis of the spherical coordinates is chosen to pass through the point at which the ray is tangent to the sphere of radius $r_\circ$ then the path of the ray in spherical coordinates is given by

$$\theta = \pi/2 - \vartheta = \pi/2 - \sin^{-1}(r_\circ/r). \tag{1.7.4}$$

If we expand $I(r,\vartheta)$ in a Taylor series:

$$I(r+dr,\vartheta+d\vartheta) = I(r,\vartheta) + \frac{\partial I(r,\vartheta)}{\partial r}dr + \frac{\partial I(r,\vartheta)}{\partial\vartheta}d\vartheta + \cdots, \tag{1.7.5}$$

keep only first order terms in small quantities, and substitute this and the expressions 1.7.2 into 1.7.1, we obtain the basic equation of radiative transport:

$$\frac{\partial I_\nu(r,\vartheta)}{\partial r}\cos\vartheta - \frac{\partial I_\nu(r,\vartheta)}{\partial\vartheta}\frac{\sin\vartheta}{r} + \kappa_\nu\rho I_\nu(r,\vartheta) - \frac{j_\nu\rho}{4\pi} = 0. \tag{1.7.6}$$

The subscript ν denotes the dependence of I, κ, and j on photon frequency; properly I_ν and j_ν are defined per unit frequency interval. Henceforth we do not make this subscript or the arguments (r, ϑ) explicit unless they are being discussed.

We are usually more interested in quantities like the energy density of the radiation field and the rate at which it transports energy than in the full dependence of I on angle. Fortunately, these quantities may be represented as angular integrals over I, and are intrinsically much simpler quantities which satisfy much simpler equations than (1.7.6). Only in the very detailed study of stellar atmospheres is the full angular dependence of I significant. The following quantities are important:

$$\frac{4\pi}{c} J \equiv \mathcal{E}_{rad} \equiv \frac{1}{c} \int I \, d\Omega \tag{1.7.7a}$$

$$H \equiv \int I \cos\vartheta \, d\Omega \tag{1.7.7b}$$

$$\frac{4\pi}{c} K \equiv P_{rad} \equiv \frac{1}{c} \int I \cos^2\vartheta \, d\Omega . \tag{1.7.7c}$$

In (1.7.7a) and (1.7.7c) two symbols have been defined because both are in common use. Sometimes H is defined as $\frac{1}{4\pi}$ times the definition in (1.7.7b). The integrals in (1.7.7) are called the angular moments of I; clearly an infinite number of such moments may be defined, but these three are usually the only important ones. It is evident that $\mathcal{E}_{rad}$ is the energy density of the radiation field, H is the radiation flux (the rate at which radiation carries energy across a unit surface normal to the $\vartheta = 0$ direction), and P_{rad} is the radiation pressure. As defined these quantities are functions of frequency, but formally identical relations apply to their integrals over frequency.

In general the n-th moment (where n is the power of $\cos\vartheta$ appearing in the integrand) is a tensor of rank n; the scalar expressions of (1.7.7b) and (1.7.7c) refer to the z component of the flux vector and the zz component of the radiation stress tensor, where $\hat{z}$ is the unit vector along the $\vartheta = 0$ axis. In practice, the z component of H is usually the only nonzero one and the stress tensor is usually nearly isotropic so that it may be described by a scalar P_{rad}.

It is now easy to obtain differential equations for the simpler quantities $\mathcal{E}_{rad}$, H, P_{rad} by taking angular moments of equation (1.7.6); that is, by applying $\int \cos^n \vartheta \, d\Omega$ to the entire equation and carrying out the integrals. The zeroth and first moments are

$$\frac{dH}{dr} + \frac{2}{r}H + c\kappa\rho\mathcal{E}_{rad} - j\rho = 0 \qquad (1.7.8a)$$

$$\frac{dP_{rad}}{dr} + \frac{1}{r}(3P_{rad} - \mathcal{E}_{rad}) + \frac{\kappa\rho}{c}H = 0. \qquad (1.7.8b)$$

There is an evident problem with this procedure: we have two equations for the three quantities $\mathcal{E}_{rad}$, H, and P_{rad}. If we obtain a third equation by taking the second moment of (1.7.6) we must evaluate integrals like $\int I \cos^3 \vartheta \, d\Omega$, which introduce a fourth quantity, the third moment of I. It is evident that this problem will not be solved exactly by taking any finite number of moments; it arises very generally in moment expansions in physics.

In practice moment expansions are truncated; only a small finite number of moments are taken, and some other information, usually approximate, is used to supply the missing equation. In order to do this expand I in a power series in $\cos\vartheta$:

$$I = I_0 + I_1 \cos\vartheta + I_2 \cos^2\vartheta + \cdots. \qquad (1.7.9)$$

We could also expand in Legendre polynomials, which would have the advantage of being orthogonal functions, but for the argument to be made here this is unnecessary. Substitute this power series into (1.7.6), and equate the coefficients of each power of ϑ in the resulting expression to zero. There results an infinite series of algebraic equations whose first three members are:

$$\frac{I_1}{r} + \kappa\rho I_0 = \frac{j\rho}{4\pi} \tag{1.7.10a}$$

$$\frac{\partial I_0}{\partial r} + \frac{2I_2}{r} + \kappa\rho I_1 = 0 \tag{1.7.10b}$$

$$\frac{\partial I_1}{\partial r} - \frac{I_1}{r} + \frac{3I_3}{r} + \kappa\rho I_2 = 0. \tag{1.7.10c}$$

We now need only to estimate the order of magnitude of the I_n, so we may replace $\frac{\partial}{\partial r}$ by $1/l$ and r by l where l is a characteristic length (noting that $\frac{\partial}{\partial r}$ and $-1/r$ do not cancel because this is only an order-of-magnitude replacement—instead, their sum is still of order $1/l$). Again, we have one more variable than equations. However, these equations have an approximate solution for which terms involving the extra variable become insignificant. This solution is

$$I_0 \approx \frac{j}{4\pi\kappa} \tag{1.7.11a}$$

$$I_n \sim I_0(\kappa\rho l)^{-n} \qquad n \geq 1. \tag{1.7.11b}$$

The factor $(\kappa\rho l)$ is generally very large ($\sim 10^{10}$ in the Solar interior) so the higher terms in (1.7.9) become small exceedingly rapidly. As a result (1.7.11a) holds very accurately, while (1.7.11b) is only an order of magnitude expression. It is evident that the terms in (1.7.10) which bring in more variables than equations (those of the form nI_n/r) are smaller than the other terms by a factor of order $(\kappa\rho l)^{-2}$ and are completely insignificant. (1.7.11b) is a rough approximation only because of the replacement of $\frac{\partial}{\partial r}$ by $1/l$, not because of the neglect of the terms of the form nI_n/r.

Because of (1.7.11b), (1.7.9) may be truncated after the $n = 1$ term, and $\mathcal{E}_{rad}$, H, and P_{rad} expressed to high accuracy in terms of I_0 and I_1 alone, reducing the three variables to two. The important result is that

$$P_{rad} = \frac{4\pi}{3c} I_0 = \frac{1}{3}\mathcal{E}_{rad}. \tag{1.7.12}$$

This relation between P_{rad} and $\mathcal{E}_{rad}$ is known as the Eddington approximation. By relating two of the moments of the radiation field it "closes" the moment expansion (1.7.8). It holds to high accuracy everywhere except in stellar atmospheres (in which $\kappa\rho l \sim 1$).

It might be thought that more accurate results could be obtained by taking more terms in the moment expansions. In stellar interiors this is unnecessary. Where (1.7.12) is not accurate, taking higher terms does not lead to rapid improvement. Expansions which do not converge rapidly often do not converge at all. A numerical description of the full ϑ dependence of I is a better approach.

The form of (1.7.12) is no surprise; it expresses the relation between radiation pressure and energy density in thermodynamic equilibrium, which should hold deep in a stellar interior. Similarly, if the matter at any point is locally in thermal equilibrium and there are no photon scattering processes the right hand side of (1.7.11a) equals (by the condition of detailed-balance) the black-body radiation spectrum (also called the Planck function) B_ν:

$$\frac{j_\nu}{4\pi\kappa_\nu} = B_\nu = \frac{2h\nu^3}{c^2}\frac{1}{\exp(h\nu/k_B T) - 1}. \tag{1.7.13}$$

The condition that the matter is in local thermal equilibrium (abbreviated LTE) holds to high accuracy in stellar interiors. It may fail in stellar atmospheres where the radiation field is strongly anisotropic, being mostly directed upward; such a radiation field is not in equilibrium (the Planck function is isotropic), and may drive populations of atomic levels away from equilibrium. This often produces observable effects in stellar spectra, but does not have significant effects on the gross energetics of radiative energy flow.

Scattering presents a different problem. It is simple enough to include scattering out of the beam in the opacity κ, but the source term j is more difficult, because radiation is scattered *into* the beam from all other directions (and, in some cases, from other frequencies).

In general, a term of the form

$$\int d\Omega' d\nu' \frac{d\sigma(\Omega, \Omega', \nu, \nu')}{d\Omega'} I(\Omega', \nu') \tag{1.7.14}$$

must be added to j_ν in (1.7.6), where σ is the scattering cross-section, and the solid angles Ω and Ω' describe the pairs of angles (ϑ, φ) and (ϑ', φ'). The azimuthal angles must be included to completely describe the geometry of scattering. This term is complicated; worse, it turns the relatively simple differential equation (1.7.6) into an integral equation which is much harder to solve. If the radiation field equals the Planck function, as is accurately the case in stellar interiors, then the relation (1.7.13) holds even in the presence of scattering, and it is not necessary to consider the messy integral (1.7.14).

In stellar interiors we may use the Eddington approximation (1.7.12) to reduce equations (1.7.8) to the form

$$\frac{d(Hr^2)}{dr} + c\kappa\rho\mathcal{E}_{rad} - j\rho = 0 \tag{1.7.15a}$$

$$H + \frac{c}{3\kappa\rho}\frac{d\mathcal{E}_{rad}}{dr} = 0. \tag{1.7.15b}$$

1.7.2 Spectral Averaging and Energy Flow In stellar interiors we are concerned with the flow of energy, and not with its detailed frequency dependence. We therefore wish to consider frequency integrals of our previous results. Define the luminosity $L \equiv \int 4\pi r^2 H_\nu \, d\nu$, and note that in steady state there is no net exchange of energy between the radiation and the matter, so that $\int j_\nu d\nu = \int c\kappa_\nu \mathcal{E}_{rad\nu} d\nu$. Then (1.7.15a) states that L is independent of r. For a star in steady state (as we have assumed) this is just the conservation of energy. In discussing radiative transport we have neglected nuclear energy generation; if it were included we would obtain (1.3.3).

It is more interesting to integrate (1.7.15b) over frequency. Define $H_{av} \equiv \int H_\nu d\nu$ and $\mathcal{E}_{av} \equiv \int \mathcal{E}_{rad\nu} d\nu$ so that

$$
\begin{aligned}
H_{av} &= -\frac{c}{3\rho} \int \frac{1}{\kappa_\nu} \frac{d\mathcal{E}_{rad\nu}}{dr} d\nu \\
&= -\frac{c}{3\rho} \frac{d\mathcal{E}_{av}}{dr} \frac{\int \frac{1}{\kappa_\nu} \frac{d\mathcal{E}_{rad\nu}}{dr} d\nu}{\int \frac{d\mathcal{E}_{rad\nu}}{dr} d\nu} .
\end{aligned}
\tag{1.7.16}
$$

Because the radiation field I_ν is very close to that of a black body B_ν we may write $\mathcal{E}_{rad\nu} = \frac{4\pi}{c} B_\nu$. Then (1.7.16) may be written in the simple form

$$
H_{av} = -\frac{c}{3\kappa_R \rho} \frac{d\mathcal{E}_{av}}{dr}, \tag{1.7.17}
$$

where we have defined the Rosseland mean opacity

$$
\kappa_R \equiv \frac{\int \frac{dB_\nu}{dr} d\nu}{\int \frac{1}{\kappa_\nu} \frac{dB_\nu}{dr} d\nu} = \frac{\int \frac{dB_\nu}{dT} d\nu}{\int \frac{1}{\kappa_\nu} \frac{dB_\nu}{dT} d\nu} . \tag{1.7.18}
$$

These integrals may be computed from the atomic properties of the matter and the Planck function.

The Rosseland mean κ_R is a harmonic mean, and therefore is sensitive to any "windows" (frequencies at which κ_ν is small), but is insensitive to spectral lines at which κ_ν is large. This behavior is very different from that of the frequency-integrated microscopic emissivity of matter (which gives the power radiated by low density matter for which absorption in unimportant); this emissivity is proportional to the arithmetic mean of κ_ν so that lines are important but windows are not. The spectrum of matter usually contains many absorption lines, but not windows, because there generally are processes which provide some absorption across very broad ranges of frequency. The Rosseland mean is therefore not very sensitive to uncertainties in

κ_ν, which is fortunate, because κ_ν is hard to calculate accurately. Because of the frequency dependences of $\frac{dB_\nu}{dT}$ and of typical κ_ν, κ_R is most sensitive to the values of κ_ν at frequencies for which $\frac{h\nu}{k_B T} \sim$ 3–10.

From (1.7.17) we obtain

$$H_{av} = -\frac{c}{\kappa_R \rho}\frac{dP_r}{dr}, \tag{1.7.19}$$

where P_r is the frequency-integrated radiation pressure. This relates the rate at which radiation carries energy to the gradient of radiation pressure. If the black body relation $P_r = \frac{a}{3}T^4$ is substituted in (1.7.19) and the definition of L is used then (1.3.4) is obtained.

In general $0 > \frac{dP_r}{dr} \geq \frac{dP}{dr}$ (unless the gas pressure were to *increase* outward, an unlikely event which would require that the density also increase outward, an unstable situation; see **1.8.1**). The equation of hydrostatic equilibrium (1.3.1) gives $\frac{dP}{dr}$, so that (1.7.19) implies an upper bound on H_{av} and on L for a star in hydrostatic equilibrium. This is the origin of the Eddington limit on stellar luminosities L_E used in **1.4.3**.

1.7.3 Scattering Atmospheres An interesting application of these equations is to the problem of an atmosphere in which the opacity is predominantly frequency-conserving scattering, rather than absorption. This is a good approximation for hot luminous stars, X-ray sources, and the hotter parts of accretion discs, but also for visible radiation in very cool stellar and planetary atmospheres. Define the single-scattering albedo ϖ of the material as the fraction of the opacity attributable to scattering; then $1 - \varpi \ll 1$ is the fraction attributable to absorption.

Begin with equations (1.7.8), assume a nearly isotropic radiation field and the Eddington approximation (1.7.12), and consider the case of a plane-parallel atmosphere of uniform temperature, so that $\frac{1}{r} \ll \frac{d}{dr}$ and B is independent of space. Equations (1.7.8) become

$$\frac{dH}{dr} + c\kappa\rho\mathcal{E}_{rad} - j\rho = 0 \tag{1.7.20a}$$

$$\frac{dP_{rad}}{dr} + \frac{\kappa\rho}{c}H = 0. \tag{1.7.20b}$$

The source term j is now given by

$$j = 4\pi\kappa B(1-\varpi) + \kappa\mathcal{E}_{rad}c\varpi; \tag{1.7.21}$$

substitution leads to

$$\frac{1}{\kappa\rho}\frac{dH}{dr} + \mathcal{E}_{rad}c(1-\varpi) - 4\pi B(1-\varpi) = 0. \tag{1.7.22}$$

Define the optical depth τ by

$$d\tau \equiv -\kappa\rho dr, \tag{1.7.23}$$

with $\tau = 0$ outside the atmosphere (above essentially all its material); this definition is used in all radiative transfer problems. Equations (1.7.22) and (1.7.20b) become

$$\frac{dH}{d\tau} = (\mathcal{E}_{rad}c - 4\pi B)(1-\varpi) \tag{1.7.24a}$$

$$\frac{1}{3}\frac{d\mathcal{E}}{d\tau} = \frac{H}{c}. \tag{1.7.24b}$$

Differentiation of (1.7.24b) and substitution into (1.7.24a) leads to

$$\frac{d^2(\mathcal{E}_{rad} - 4\pi B/c)}{d\tau^2} = 3(1-\varpi)(\mathcal{E}_{rad} - 4\pi B/c). \tag{1.7.25}$$

Applying the boundary condition that $\mathcal{E}_{rad} \rightarrow 4\pi B/c$ as $\tau \rightarrow \infty$ leads to the solution

$$\mathcal{E}_{rad} = \frac{4\pi B}{c}\Big[1 - \exp\big(-\sqrt{3(1-\varpi)}\tau\big)\Big]. \tag{1.7.26}$$

One consequence of this result is that the radiation field does not approach the black body radiation field until $\tau \gtrsim [3(1-\varpi)]^{-1/2} \gg 1$; in an atmosphere with largely absorptive opacity the corresponding condition is $\tau \gtrsim 1$.

Another consequence is found when we compute the emergent radiant power $H(\tau = 0)$ from (1.7.24b):

$$H = \frac{4\pi B}{3}\sqrt{3(1-\varpi)}. \tag{1.7.27}$$

This should be compared to the result for a black body radiator $H = \pi B$, which is obtained from (1.7.7b) if $I = B$ for $\vartheta \leq \pi/2$, and $I = 0$ for $\vartheta > \pi/2$. The scattering atmosphere radiates a factor of $\frac{4}{3}\sqrt{3(1-\varpi)} \ll 1$ as much power as a black body at the same temperature. This may be described as an emissivity $\varsigma = \frac{4}{3}\sqrt{3(1-\varpi)} \ll 1$ of the scattering atmosphere; by the condition of detailed balance such an atmosphere has an angle-averaged albedo (the fraction of incident flux returned to space after one or more scatterings) of $1 - \varsigma$. If it has an effective temperature T_e, its actual temperature $T \approx \varsigma^{-1/4} T_e \approx 0.81(1-\varpi)^{-1/8} T_e$, where we have assumed that ϖ and ς are not strongly frequency dependent.

The high albedo of a medium whose opacity is mostly scattering is observed in everyday life when one adds cream to coffee. The extract of coffee we drink is a nearly homogeneous substance whose opacity is almost entirely absorptive; its albedo is very low. The mixture of coffee and cream is visibly lighter in appearance because of the high scattering cross-sections of globules of milk fat. The reduced emissivity of the mixture is unobservable, because the Planck function is infinitesimal at visible wavelengths and room temperature.

Equation (1.7.27) appears to imply $\varsigma > 1$ if $\varpi \to 0$, but this thermodynamically impossible result is incorrect because the assumption of the Eddington approximation is invalid for $\tau \lesssim 1$, which is the important region in determining the emergent flux from an absorbing atmosphere. In a scattering atmosphere, optical depths up to $[3(1-\varpi)]^{-1/2} \gg 1$ are important; the Eddington approximation is valid over most of this range.

1.8 Turbulent Convection

If we heat the bottom and cool the top of a reservoir of fluid at rest, heat will flow upward. The central regions of stars are heated by thermonuclear reactions and their surfaces are cooled by radiation. If the rate of heat flow is low, it will flow by a combination of radiation and conduction. Conduction is usually dominant in everyday liquids and in degenerate stellar material, and radiation is usually dominant in gases, at high temperatures, and in nondegenerate stellar interiors. At high heat fluxes a new process appears, in which macroscopic fluid motions transport warmer material upward and cooler material downward. This process is called convection. For limited parameter ranges convection may take the form of a laminar flow, but in astronomy it is almost always turbulent, if it occurs at all. We must ask when it occurs and what are its consequences.

1.8.1 Criteria Two criteria must be satisfied in order to have convection. The first is that viscosity not be large enough to prevent it. This is an important effect in small laboratory systems, and successful quantitative theories exist, but in stellar heat transport the influence of viscosity is negligible; if convection takes place at all Reynolds numbers usually exceed 10^{10}.

The more important criterion is that the thermodynamic state of the stellar interior be such that convective motions release energy, rather than requiring energy to drive them. In other words, convection will occur if it carries heat from hotter regions to cooler ones (given the well-justified assumption that viscosity is a negligible retarding force), but not if it were to carry heat from cooler regions to hotter ones.

To make this criterion more quantitative we compare the thermodyamic state of the star at two radii separated by a small radius increment dr; at r_l the pressure is P_l and the density is ρ_l, while at r_u the pressure is P_u and the density is ρ_u. We assume that the chemical composition is uniform and that densities and opacities are high enough that radiative transport of energy is negligible on the timescales of convective motions; these assumptions are usually (but not always) justified in stellar interiors, but fail in stellar atmospheres. We also relate adiabatic variations in the pressure and density of the fluid by an equation of state of the form

$$P \propto \rho^{\gamma}. \tag{1.8.1}$$

Such a fluid is known as a "γ-law" gas; γ is discussed in **1.9.1** and is usually between 4/3 and 5/3. It is here only necessary to assume that the form (1.8.1) holds for adiabatic processes over small ranges of P and ρ; this will be the case for any fluid except near a phase transition.

Now consider raising an element of fluid from the lower level to the upper one, with all fluid velocities slow (much slower than the sound speed) so that the fluid element remains in hydrostatic equilibrium with its mean surroundings. When it reaches the upper level it has a density ρ'_u given by

$$\rho'_u = \rho_l \left(\frac{P_u}{P_l}\right)^{\frac{1}{\gamma}} \approx \rho_l \left(1 + \frac{1}{\gamma P_l}\frac{dP}{dr}dr\right). \tag{1.8.2}$$

If $\rho'_u > \rho_u$ then the raised fluid element is denser than its surroundings and will tend to fall back to its initial position. In this case the fluid is stable against convective displacement. A more quantitative analysis would calculate the frequency of sinusoidal perturbations of the horizontal fluid layers (analogous to water surface waves, but allowing for the continuous variation of P and ρ), and would find their frequency to be real.

If $\rho'_u < \rho_u$ the raised fluid is less dense than its surroundings, and experiences a further buoyancy force which accelerates its rise. A similar calculation of the density of a fluid element descending from the upper layer shows that for it $\rho'_l > \rho_l$, so negative buoyancy accelerates its descent. In this case the fluid is unstable, and convective motions begin. In the more quantitative analysis the perturbations of the horizontally layered structure have imaginary frequencies of both signs, and grow exponentially.

For small dr we may write $\rho_u \approx \rho_l + \frac{d\rho}{dr} dr$ so that the stability condition becomes

$$-\frac{1}{\gamma P}\frac{dP}{dr} < -\frac{1}{\rho}\frac{d\rho}{dr}. \tag{1.8.3}$$

This awkward-appearing form with minus signs on each side has been chosen because the derivatives are both negative.

The definition of an incompressible fluid is that $\gamma \to \infty$; then the stability criterion (1.8.3) becomes $\frac{d\rho}{dr} < 0$, a familiar result. It is apparent that for compressible fluids as well $\frac{d\rho}{dr} > 0$ would make stability impossible (because the equation 1.3.1 of hydrostatic equilibrium requires $\frac{dP}{dr} < 0$). For an adiabatic equation of state of the form (1.8.1) the entropy $S \propto \ln(P/\rho^\gamma)$, and the stability condition takes the form

$$0 < \frac{dS}{dr}. \tag{1.8.4}$$

These stability conditions are local; it is clear that if an unstable interchange is possible between two widely separated layers (1.8.3) and (1.8.4) will be violated for at least a portion of the region between the layers.

The bound (1.8.3) may be transformed into a bound on $\frac{dT}{dr}$ by use of (1.4.5); the result is messy unless one of the terms in (1.4.5) is negligible. More generally, if $P \propto \rho^\alpha T^\beta$ (in contrast to 1.8.1, this refers to the functional form of $P(\rho, T)$, and *not* to its variation under adiabatic processes) we can readily obtain

$$-\left(1 - \frac{\alpha}{\gamma}\right) \frac{1}{P}\frac{dP}{dr} > -\frac{\beta}{T}\frac{dT}{dr}. \tag{1.8.5}$$

This is known as the Schwarzschild criterion for stability.

In this derivation we have assumed uniform chemical composition and have ignored angular momentum. Either of these may make the problem much more difficult. For example, if the matter in layer l has higher molecular weight than that in layer u this will tend to stabilize the fluid against convection. A more subtle process called semi-convection may still occur even when ordinary convection does not; it depends on the ability of energy to flow radiatively out of the denser fluid, and thus to separate itself from the stabilizing influence of the higher molecular weight. Semi-convection is one of a large class of "double-diffusive" and "multi-diffusive" processes known to astrophysicists and geophysicists.

The criterion (1.8.5) shows that there is instability when $\left|\frac{dT}{dr}\right|$ is large, and (1.3.4) shows that this tends to occur when κ or L/r^2 are large. Detailed calculations show that (1.8.3–5) are violated in the outer layers of stars with cool surfaces (including the Sun) because at low temperatures κ is large, and near the energy-producing regions of luminous stars, where L/r^2 is large.

1.8.2 Consequences Suppose (1.8.3–5) are violated; what then? It is clear that the interchange of elements of matter which are unstable against interchange will tend to reduce ρ_u and to increase ρ_l, and to increase S_u and to decrease S_l. The limiting state of this process is to turn the violated inequalities (1.8.3–5) into equalities whch then describe the variation of P, ρ, T, and S in the star. Any one of these equalities (they are all equivalent) then replaces (1.3.4) in describing the thermal structure of the star. In other words, the effect of convective instability is to eliminate the conditions which gave rise to it. This is a natural and plausible hypothesis which is widely assumed in turbulent flow problems. It cannot be exactly true; some small excess $\left|\frac{dT}{dr}\right|$ must remain to drive the convective flow.

A crude argument exists to estimate the accuracy of this approximation; the estimate is based on an adaptation of Prandtl's mixing length theory of turbulent flows. Although reality is surely more complex, imagine that the turbulent flow is composed of discrete fluid elements which rise or fall without drag forces (but remain in pressure equilibrium with their surroundings) for a distance ℓ from their origins. After travelling this distance they mix with their new surroundings and lose their identity. Denote the excess of the temperature gradient over the value given by (1.8.5) (taken as an equality) by $\Delta\nabla T$; it is this quantity (called the superadiabatic temperature gradient) we must estimate. After a rising fluid element has travelled a distance dr its temperature exceeds that of its mean surroundings by an amount $\Delta\nabla T dr$; its own thermodynamic state has varied exactly adiabatically and it remains in pressure equilibrium with its mean surroundings (both by assumption). A falling fluid element is similarly cooler than its mean surroundings by $\Delta\nabla T dr$. The combination of rising warmer fluid and falling cooler fluid produces a mean convective heat flux

$$H_{conv} \sim \Delta\nabla T dr c_P \rho v, \tag{1.8.6}$$

where v is a typical flow velocity and c_P is the specific heat at constant P.

In order to estimate v we use the assumption that the only forces acting on fluid elements are those of buoyancy. We have

$$\frac{\Delta\nabla\rho}{\rho} = \left(\frac{\beta}{\gamma - \alpha}\right)\frac{\Delta\nabla T}{T} \sim \frac{\rho}{T}\Delta\nabla T, \tag{1.8.7}$$

and the buoyancy force (which is proportional to dr) leads to a velocity

$$v^2 = \frac{GM(r)}{r^2}\frac{\Delta\nabla\rho}{\rho}(dr)^2 \sim \frac{GM(r)}{r^2}\frac{\Delta\nabla T}{T}(dr)^2. \tag{1.8.8}$$

Now evaluate these expressions after fluid elements have travelled half of the mixing length, so that $dr = \ell/2$:

$$H_{conv} \sim \frac{c_P\rho\ell^2}{4}\sqrt{\frac{GM(r)}{r^2T}}(\Delta\nabla T)^{3/2}. \tag{1.8.9}$$

A sensible choice of ℓ is a matter of guesswork; it is usually taken to be comparable to the pressure scale height $\left|\frac{d\ln P}{dr}\right|^{-1}$. Observations of the Solar surface show that the convective motions are very complex. The visible surface is divided into a network of small polygonal cells, called granules, which are columns of rising fluid bounded by regions of descending fluid. There is also a larger scale pattern of supergranulation. These observations do not provide direct evidence concerning the vertical mixing length, and flows in the observable Solar atmosphere (where the scale height is small) may not resemble those in deeper layers.

If ℓ is the pressure scale height and $H_{conv} = L/(4\pi r^2) - H_{av}$ (where H_{av} is the radiative flux calculated in **1.7**) then we can evaluate $\Delta\nabla T$ and v at various places in a star. Our results may be manipulated to yield

$$\Delta\nabla T \sim \left|\frac{dT}{dr}\right|\left(\frac{\ell}{r}\right)^{-4/3}\left(\frac{t_h}{t_{th}}\right)^{2/3} \tag{1.8.10a}$$

$$v^2 \sim c_s^2 \left(\frac{T_c}{T}\right) \left(\frac{\ell}{r}\right)^{2/3} \left(\frac{t_h}{t_{th}}\right)^{2/3} \tag{1.8.10b}$$

where the thermal time t_{th} has been redefined (from 1.6.2) to include only the thermal energy content of the convective region, T_c is the central temperature, and c_s is the sound speed. For the convective regions of the Sun (but not its surface layers) $\Delta\nabla T \sim 10^{-6}\left|\frac{dT}{dr}\right|$ and $v \sim 10^{-4}c_s \sim 30$ m/sec. Thus the adiabatic approximation to the structure of a convective zone—the adoption of (1.8.3–5) as equalities—is usually justified to high accuracy, even though the estimates (1.8.6–9) are very crude. Similarly, characteristic hydrodynamic stresses are $\sim \rho v^2 \sim 10^{-8}P$, which establishes that the assumption that fluid elements remain in hydrostatic equilibrium also holds to high accuracy. The time for fluid to circulate through the Solar convective region is $\sim \ell/v \sim 1$ month, which is short enough to guarantee complete mixing.

These approximations break down in the surface layers of stars, as shown by equations (1.8.10). In these layers the scale height and ℓ become small, as do ρ, T, and t_{th} ($t_{th} \approx c_P \rho T \ell / H$). It is not possible to calculate quantitatively the structure of these layers. This problem is most severe for cool giants and supergiants, where T and especially ρ become very small. Their surfaces may not be spherical or in hydrostatic equilibrium, but may rather consist of geysers or fountains of gas which erupts, radiatively cools, and then falls back.

It is important to realize that H_{conv} (1.8.9) is not directly related to or limited by the pressure gradient, unlike the radiative H_{av} (1.7.17). This means that in stellar interiors convection may carry a nearly arbitrarily large luminosity, and the Eddington limit L_E does not apply.

Near stellar surfaces this problem is more complicated because there $\Delta\nabla T$ becomes large for large H_{conv}. In the low densities of stellar atmospheres convection is incapable of carrying a large heat flux because the thermal energy content of the matter is low, and energy must flow by radiation. For hot stars the opacity is essentially constant and radiative transport in the upper atmosphere imposes the upper bound L_E on the stellar luminosity. For cool giants and supergiants the opacity in the upper atmosphere may be extremely small, and no simple bound on the luminosity exists. The actual luminosity of fully convective stars is determined by these surface layers in which the approximation of nearly adiabatic convection breaks down, and no satisfactory theory exists.

1.9 Constitutive Relations

Each of the constitutive relations (1.3.5–7) is an extensive field of research which extends far beyond the scope of this book. This section presents only the sketchiest overview of a few qualitative conclusions which should be familiar to every astrophysicist.

1.9.1 Adiabatic Exponent Here we derive a few useful results. Because stars are large and opaque, and t_{th} is usually long, we are often concerned with the properties of matter undergoing adiabatic processes.

Consider a perfect gas which satisfies the equation of state (1.4.5)

$$P = \frac{\rho N_A k_B T}{\mu} \tag{1.9.1}$$

where we now neglect radiation pressure. For a gram of gas undergoing a reversible process

$$dQ = d\mathcal{U} + P dV \tag{1.9.2}$$

where dQ is an infinitesimal increment of heat, $\mathcal{U}(V, T)$ is the internal energy per gram, and $V \equiv 1/\rho$ is the volume per gram. We define a perfect gas by the condition that $\mathcal{U}$ depend only on T: $\mathcal{U}(V,T) = \mathcal{U}(T)$.

The specific heats at constant pressure and at constant volume, c_P and c_V respectively, are defined:

$$c_P \equiv \left.\frac{dQ}{dT}\right|_P \tag{1.9.3a}$$

$$c_V \equiv \left.\frac{dQ}{dT}\right|_V, \tag{1.9.3b}$$

where the subscript denotes the thermodynamic variable to be held constant. From (1.9.2), using (1.9.1) to eliminate P

$$c_V = \frac{d\mathcal{U}}{dT} \tag{1.9.4a}$$

$$c_P = \frac{d\mathcal{U}}{dT} + \frac{N_A k_B}{\mu}. \tag{1.9.4b}$$

The definition of an adiabatic process is that $dQ = 0$. From the preceding equations and definitions we find for such a process

$$0 = c_V dT + (c_P - c_V)\frac{T}{V}dV. \tag{1.9.5}$$

Defining $\gamma \equiv c_P/c_V$ yields

$$0 = d\ln T + (\gamma - 1)d\ln V. \tag{1.9.6}$$

Integrating this equation, using the definition of V and (1.9.1), yields

$$P \propto \rho^\gamma. \tag{1.9.7}$$

The ratio of specific heats depends on the atoms or molecules making up the gas. By explicit calculation of $\mathcal{U}$ for a perfect gas it is easy to see that

$$\gamma = \frac{q+2}{q} \tag{1.9.8}$$

where q is the number of degrees of freedom excited per atom or molecule. For a monatomic gas $q = 3$, for a diatomic gas in which the vibrational degrees of freedom are not excited (such as air under ordinary conditions) $q = 5$, while for a gas of large molecules or one undergoing temperature-sensitive dissociation or ionization $q \to \infty$. In stellar interiors we may usually take $q = 3$ and $\gamma = 5/3$, except in regions of partial ionization or where radiation pressure or relativistic degeneracy are important.

In this simple derivation it was necessary to assume a perfect gas and to exclude radiation pressure. These may be included, but lead to much more complex results. For a gas consisting only of radiation this derivation is invalid because $c_P \to \infty$; T is a unique function of P so that at fixed P no amount of added energy can raise the temperature.

From the relation (1.9.7) describing adiabatic processes we can derive a relation between P and the internal energy per volume $\mathcal{E}$. Taking logarithmic derivatives of (1.9.7) and using the definition of V we obtain

$$V dP = -\gamma P dV. \tag{1.9.9}$$

Adding PdV to each side gives

$$V dP + P dV = -(\gamma - 1) P dV \tag{1.9.10a}$$

$$d\left(\frac{PV}{\gamma - 1}\right) = -P dV. \tag{1.9.10b}$$

In an adiabatic process the work done by the fluid on the outside world is $-PdV$, so that (1.9.10b) has the form of a condition of conservation of energy for the fluid, with the left hand side being the increment in internal energy. Then the internal energy per unit volume $\mathcal{E}$ is given by

$$\mathcal{E} = \frac{P}{\gamma - 1}. \tag{1.9.11}$$

The order of the manipulations between (1.9.7) and (1.9.11) may be reversed, so that these two relations are equivalent.

It is important to note that the equivalence between (1.9.7) and (1.9.11) does not require the assumption of a perfect gas or the definition of the specific heats, so that it applies even where it is not possible to derive γ as a ratio of specific heats. The most important application of this is to radiation. From (1.7.12) (or 1.7.7), for a black body radiation field $\mathcal{E}_{rad} = 3P_{rad}$, so that $\gamma = 4/3$ and (1.9.7) describes adiabatic processes in a gas of equilibrium radiation.

1.9.2 Degeneracy The matter in degenerate dwarves, the cores of some giant and supergiant stars, and in neutron stars is Fermi-degenerate. By this we mean that the thermal energy $k_B T$ is much less than the Fermi energy ϵ_F (or, more properly, the chemical potential of the degenerate species), so that states with energies up to ϵ_F are nearly all occupied, and those with higher energies are nearly all empty. This resembles the familiar metallic state of matter. The degenerate species is usually the electron; in neutron stars free neutrons are also degenerate, hence their name.

The density n_d of the degenerate fermion species is given by

$$n_d = 2\left(\frac{4}{3}\pi p_F^3\right)\frac{1}{h^3}, \qquad (1.9.12)$$

where p_F is the momentum corresponding to the Fermi energy ϵ_F. This is a standard result of elementary statistical mechanics, obtained by counting volumes in phase space, or by calculating the eigenstates of free particles in a box. The factor of 2 comes from the statistical weight of spin 1/2 particles.

For noninteracting nonrelativistic particles of mass m_d we have

$$\epsilon_F = \frac{p_F^2}{2m_d} \propto n_d^{2/3}, \qquad (1.9.13)$$

while characteristic Coulomb energies vary with density as $\epsilon_C \propto e^2 n_d^{1/3}$. Thus at high densities $\epsilon_F \gg \epsilon_C$ and degenerate electrons may be accurately treated as non-interacting particles. This makes the calculation of their equation of state easy and accurate, because the complex band structure of ordinary metals (for which $\epsilon_F \sim \epsilon_C$) may be neglected. The cohesion of ordinary metals (the fact that they have $P = 0$ at finite n_d) requires that ϵ_C be comparable to ϵ_F.

The pressure and internal energy of noninteracting degenerate nonrelativistic particles are found by integrating over their distribution function:

$$\begin{aligned} P &= \int_0^{p_F} p_x v_x \frac{2}{h^3} d^3p \\ &= \frac{1}{3}\int_0^{p_F} m_d v^2 \frac{2}{h^3} d^3p \\ &= \frac{8\pi p_F^5}{15 m_d h^3} \\ &\propto \rho^{5/3} \end{aligned} \tag{1.9.14a}$$

$$\begin{aligned} \mathcal{E} &= \int_0^{p_F} \frac{m_d v^2}{2} \frac{2}{h^3} d^3p \\ &= \frac{3}{2}P, \end{aligned} \tag{1.9.14b}$$

where we have used the fact that $\langle p_x v_x \rangle = \frac{1}{3}\langle p_x v_x + p_y v_y + p_z v_z \rangle = \frac{1}{3}\langle pv \rangle$ for a distribution function which is isotropic in 3-dimensional momentum space; here unsubscripted p and v denote their magnitudes. The relation between $\mathcal{E}$ and P, which corresponds to $\gamma = 5/3$, depends only on the fact that the particle energy $\epsilon_p = \frac{1}{2}pv$, and not on the form of the distribution function; hence it applies to all noninteracting gases of nonrelativistic particles, whether degenerate, nondegenerate, or partially degenerate ($\epsilon_F \approx k_B T$).

If the density is very high most of the particles are relativistic, $\epsilon_p \approx pc$ and $v_x \approx c p_x / p$. If we assume this relation holds exactly over the entire distribution function then

$$\begin{aligned} P &= \int_0^{p_F} \frac{p_x^2 c}{p} \frac{2}{h^3} d^3p \\ &= \frac{1}{3} \int_0^{p_F} pc \frac{2}{h^3} d^3p \\ &= \frac{2\pi c p_F^4}{3h^3} \\ &\propto \rho^{4/3} \end{aligned} \tag{1.9.15a}$$

$$\begin{aligned} \mathcal{E} &= \int_0^{p_F} pc \frac{2}{h^3} d^3p \\ &= 3P. \end{aligned} \tag{1.9.15b}$$

The relation between $\mathcal{E}$ and P, which corresponds to $\gamma = 4/3$, depends only on the relativistic relation $\epsilon_p = pc$, and not on the form of the distribution function; hence it applies to all noninteracting relativistic gases whether degenerate or not; it even applies to bosons, which is why we recover the relation (1.7.12) for photons.

Between the nonrelativistic and relativistic limits is a regime in which neither (1.9.14) nor (1.9.15) is accurate, and $4/3 < \gamma < 5/3$. This transition occurs for $p_F \approx m_d c$, which by (1.9.12) occurs at a density

$$n_d \approx \frac{8\pi m_d^3 c^3}{3h^3}. \tag{1.9.16}$$

For degenerate electrons this corresponds to $\rho \approx 2 \times 10^6$ gm/cm^3, while for neutrons $\rho \approx 10^{16}$ gm/cm^3. These are, to order of magnitude, the characteristic densities of degenerate dwarves and neutron stars respectively.

The regions in the ρ - T plane in which various approximations to the equation of state hold are shown in Figure 1.2. Quantitative calculations exist for the intermediate cases. The regions occupied by the centers and deep interiors of ordinary stars and of degenerate dwarves are shown.

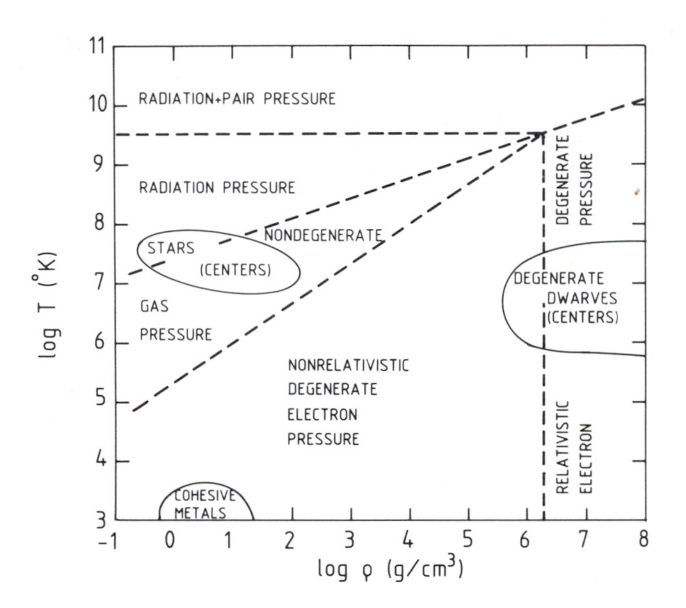

Figure 1.2. Equation of State Regimes.

The results (1.9.14) and (1.9.15) are only rough approximations for degenerate neutrons, because neutrons interact by strong nuclear forces, which are attractive at relatively large distances (several $\times 10^{-13}$ cm) but which are strongly repulsive at shorter distances.

1.9.3 Opacity A quantitative calculation of the opacity of stellar material requires elaborate calculations involving the absorption cross-sections of the ground and many excited states of many ionic species. Such calculations have been performed, and their results are available for quantitative work. It is still important to be aware of a few qualitative principles.

In all ionized matter free electrons scatter radiation, a process called Thomson or Compton scattering. For nondegenerate electrons, in the limits $h\nu \ll m_e c^2$ and $k_B T \ll m_e c^2$ the scattered radiation has the same frequency as the incident radiation, and carries no net momentum. The scattering is not isotropic, but for all $0 \leq \psi \leq \pi/2$ scattering by angles ψ and by $\pi - \psi$ is equally likely; for most purposes it may be treated as if it were isotropic. The total scattering cross-section (**2.6.3**) is $\frac{8\pi e^4}{3m_e^2 c^4} = 6.65 \times 10^{-25}$ cm^2. For matter of the usual stellar composition (70% hydrogen by mass) this produces an electron scattering opacity

$$\kappa_{es} = 0.34 \text{ cm}^2/\text{gm}. \tag{1.9.17}$$

Because this opacity is essentially independent of frequency and temperature in fully ionized matter, (1.9.17) is usually a lower bound on the Rosseland mean opacity. The only circumstances in which the opacity of stellar matter may be significantly less than this value are when it is degenerate (electron scattering is suppressed because most outgoing electron states are occupied), or when it is cool enough that most of the electrons are bound to atoms. The total opacity drops below the value given by (1.9.17) for $T \lesssim 6000$°K.

A free electron moving in the Coulomb field of an ion may absorb radiation; this process is called free-free absorption or inverse bremsstrahlung. Its quantitative calculation is rather lengthy, but a simple semiclassical result is informative. This may be obtained by using the classical expression (2.6.12) or (2.6.15) for the power radiated by an accelerated charge (an electron in the Coulomb field of the ion) to calculate the emissivity, and using the condition of detailed-balance (1.7.13) to obtain from this the opacity. The resulting cross-section per electron is proportional to $n_i v^{-1} \nu^{-3}$, where n_i is the ion density, v is the electron velocity, and ν is the photon frequency. For a typical electron v will be comparable to the thermal velocity, so $v \propto T^{1/2}$, and for a representative photon $h\nu \propto T$.

Rough numerical evaluation of the Rosseland mean leads to

$$\kappa_R \sim 10^{23} \frac{\rho}{T^{7/2}} \ \mathrm{cm}^2/\mathrm{gm}; \tag{1.9.18}$$

this expression is only approximate. The functional form of (1.9.18) is known as Kramers' law.

The photoionization of bound electrons (from both ground and excited states) produces bound-free absorption. Its frequency dependence above its energy threshold is usually similar to the ν^{-3} of free-free absorption, but the abundances of the various ions, ionization states, and excitation levels must be considered too. The resulting mean opacity roughly follows Kramers' law, and is of the same order of magnitude as that attributable to free-free absorption.

Any Kramers' law opacity is large at low temperature and high density. At high temperature or low density electron scattering is the principal opacity. The dividing line is approximately given by $T \sim 5 \times 10^6 \rho^{2/7}$ °K. At low temperatures ($T \lesssim 10000$°K) the number of free electrons becomes small and most photons have insufficient energy to ionize atoms; consequently, the opacity drops precipitously and falls below κ_{es}.

The serious user of quantitative opacity information will use the tables which have been computed, but a few further qualitative points should be made:

Because the Rosseland mean is a harmonic mean, the various contributions to the mean opacity are not additive unless they have the same frequency dependence.

Absorption opacities contain a factor $[1 - \exp(-h\nu/k_B T)]$ whose physical origin is the effect of stimulated emission. This must be included when the Rosseland mean is computed; it is implied by the factor of this form contained in B_ν in (1.7.13); LTE of the atomic and ionic levels has been assumed.

Scattering opacities do not contain a stimulated emission factor if the scattering conserves frequency. The total rate of scattering from state i to state f is proportional to $n_i(1+n_f)$, where n_i and n_f are the occupation numbers of the corresponding photon states; $n_i n_f$ is the rate of stimulated scattering. From this must be subtracted the rate $n_f(1+n_i)$ of scatterings from f to i. The net rate is proportional to $n_i - n_f$, where n_i gives the scattering rate implied by the scattering cross-section without any stimulated scattering term, and n_f gives the the scattering contribution to the source term j. The absence of an explicit stimulated scattering factor is of little importance in stellar interiors, but may be significant in laser experiments in which n_i and n_f may be very large.

Degenerate matter, like ordinary metals, is a good conductor of heat, and in it the radiative transport of energy is usually insignificant. Because the conductive heat flux is proportional to the temperature gradient, a relation like (1.3.4) may be defined in which κ includes also the effects of conduction.

1.9.4 Thermonuclear Energy Generation Many nuclear reactions are involved in the thermonuclear production of energy and the transmutation of lighter elements into heavier ones. Each presents special problems. Here I briefly discuss a few general principles. Quantitative calculation of reaction rates in stellar interiors requires more careful attention to many details; see, for example, Clayton (1968) and Harris *et al.* (1983).

The radius of a nucleus containing A nucleons is approximately given by

$$R \approx 1.4 \times 10^{-13} A^{1/3} \text{cm}. \tag{1.9.19}$$

The electrostatic energy required to bring two rigid and unpolarizable spherical nuclei of radii R_1 and R_2 and atomic numbers Z_1 and Z_2

into contact, if their charges are concentrated at their centers, is

$$E_C = \frac{Z_1 Z_2 e^2}{R_1 + R_2} \approx \frac{Z_1 Z_2}{A_1^{1/3} + A_2^{1/3}} \text{ MeV}. \tag{1.9.20}$$

Once the nuclei touch strong attractive nuclear forces take over. In the centers of main sequence stars $k_B T$ is in the range $\frac{1}{2}$ – 4 KeV so that it is evident that conquering the Coulomb barrier is the chief obstacle to thermonuclear reactions.

The Coulomb barrier is overcome by tunnelling, in a manner first calculated by Gamow; nuclei with energies much less than E_C may (infrequently) react. We work in the center-of-mass frame of the two nuclei, so that $m = \frac{M_1 M_2}{M_1 + M_2}$ is their reduced mass, r their separation, and $k = \sqrt{2mE_\circ}/\hbar$ and $E_\circ$ are the wave-vector and kinetic energy at infinite separation. The barrier tunnelling probability P_0 is calculated in the W. K. B. approximation as

$$P_0 \sim \exp\left(-2\int_R^{r_\circ} \sqrt{\frac{2me^2 Z_1 Z_2}{\hbar^2 r} - k^2}\, dr\right) \equiv \exp(-\mathcal{I}), \tag{1.9.21}$$

where we write only the very sensitive exponential term, neglecting more slowly varying factors. Here $R = R_1 + R_2$ is the separation at contact (within which the nuclear interactions make the potential attractive), $r_\circ = \frac{2me^2 Z_1 Z_2}{\hbar^2 k^2}$ is the classical turning point (at which the integrand is zero), and the subscript 0 indicates that we consider only the $l = 0$ partial wave. Higher angular momentum states produce much smaller P_l.

The exponent in (1.9.21) may be calculated:

$$\begin{aligned} \mathcal{I} &= 2k \int_R^{r_\circ} \sqrt{\frac{r_\circ}{r} - 1}\, dr \\ &= 4kr_\circ \int_{\sqrt{R/r_\circ}}^1 \sqrt{1 - \varsigma^2}\, d\varsigma, \end{aligned} \tag{1.9.22}$$

where $\varsigma \equiv \sqrt{r/r_\circ}$. Now $\sqrt{R/r_\circ} \ll 1$ so that we may expand the integral in a power series in $\sqrt{R/r_\circ}$ with the result:

$$\begin{aligned} \mathcal{I} &= 4kr_\circ \left(\int_0^1 \sqrt{1-\varsigma^2}\, d\varsigma - \int_0^{\sqrt{R/r_\circ}} 1\, d\varsigma + \cdots \right) \\ &= 4kr_\circ \left(\frac{\pi}{4} - \sqrt{\frac{R}{r_\circ}} + \cdots \right). \end{aligned} \tag{1.9.23}$$

The leading term in (1.9.23) does not depend on R at all; this is fortunate because it implies that to a good approximation the result is independent of the nuclear sizes or to the form of the potential near nuclear contact, where it is poorly known. We now have

$$\mathcal{I} = \frac{\pi Z_1 Z_2 e^2}{\hbar} \sqrt{\frac{2m}{E_\circ}} - 4\frac{e}{\hbar}\sqrt{2mZ_1 Z_2 R} + \cdots. \tag{1.9.24}$$

The second term is independent of energy; it affects the reaction rate but we do not consider it further. The third and higher terms are small. The first term is large and after exponentiation makes the reaction rate a sensitive function of $E_\circ$.

We now must average the reaction rate over the thermal equilibrium distribution of nuclear kinetic energies. When we transform variables from the velocities of the reacting nuclei to the center-of-mass and relative velocities v_{cm} and v_{rel}, we find that the kinetic energy $\frac{1}{2}M_1 v_1^2 + \frac{1}{2}M_2 v_2^2 = \frac{1}{2}(M_1 + M_2)v_{cm}^2 + \frac{1}{2}mv_{rel}^2$, so that the distribution function of the relative motion of the reduced mass m is Maxwellian at the particle temperature T. Then the total reaction rate is given by the average over the distribution function $\langle \sigma v_{rel} \rangle$, where σ is the reaction cross-section and contains the critical factor $\exp(-\mathcal{I})$. Aside from slowly varying factors this leads to

$$\langle \sigma v_{rel} \rangle \sim \int_0^\infty \exp\left(-\frac{E}{k_B T} - \frac{B}{\sqrt{E}} \right) dE, \tag{1.9.25}$$

where $B \equiv \pi Z_1 Z_2 e^2 \sqrt{2m}/\hbar$.

The first term in the exponent in (1.9.25) declines rapidly with increasing E, while the second increases rapidly. For $B^2 \gg k_B T$ (almost always the case) their sum has a fairly narrow maximum, and when exponentiated the peak is very narrow. We therefore find the maximum and expand around it. By elementary calculus

$$-\frac{E}{k_B T} - \frac{B}{\sqrt{E}} = -\frac{3E_G}{k_B T} - \frac{3}{8}\frac{B}{E_G^{5/2}}(E - E_G)^2 + \cdots, \tag{1.9.26}$$

where the Gamow energy E_G has been defined

$$E_G \equiv \left(\frac{Bk_B T}{2}\right)^{2/3}. \tag{1.9.27}$$

Now the integral in (1.9.25) may be carried out by taking only the first two terms of (1.9.26) and extending the lower limit of integration to $-\infty$, with the result

$$\begin{aligned} \langle \sigma v_{rel} \rangle &\sim \sqrt{\frac{8\pi E_G^{5/2}}{3B}} \exp\left(-\frac{3E_G}{k_B T}\right) \\ &\sim \exp\left[-3\left(\frac{\pi^2 Z_1^2 Z_2^2 e^4 m}{2\hbar^2 k_B T}\right)^{1/3}\right], \end{aligned} \tag{1.9.28}$$

where in the last expression the slowly varying factor has been dropped, as similar factors were before, leaving only the dominant exponential dependence. This result gives the dominant temperature dependence of nonresonant thermonuclear reactions.

Under typical conditions of interest the argument of the cube root in (1.9.28) is $\sim 10^4$. It is therefore apparent that P_0 and $\langle \sigma v_{rel} \rangle$ are very small, as must be the case, in order that the nuclei in a dense stellar interior survive for 10^6–10^{10} years before reacting. It is then evident that the reaction rate is a steeply increasing function of T, and a steeply decreasing function of $Z_1 Z_2$. The sensitivity to T

implies that thermonuclear energy generation acts nearly as a thermostat when in a star whose effective specific heat is negative (see **1.5**), and tends to produce rapid instability when the effective specific heat is positive (as is the case in degenerate matter or for thin shells). It also means that when energy is produced by a given nuclear reaction T is a weak function of the other parameters. The sensitivity to $Z_1 Z_2$ implies that in most circumstances the reactions which proceed most rapidly are those with the smallest product $Z_1 Z_2$.

Real nuclear physics makes the problem more complex. If the reaction of interest is resonant at near-thermal energies (as some important ones are) this may increase the reaction rates by a large factor. The peculiar properties of nuclei with $A = 2$, 5, and 8 are also worthy of note:

The only stable nucleus with $A = 2$ is the deuteron. To produce it from protons requires the reaction

$$p + p \rightarrow D + e^{+} + \nu_e. \tag{1.9.29}$$

Because this reaction depends on the weak interaction (it amounts to a β-decay from an unbound diproton state), its rate is many orders of magnitude lower than would otherwise be the case. Yet there is no other direct way of combining two protons; the diproton is not a bound nucleus at all, but is better described as a pole of the *p-p* scattering matrix. Were the diproton bound, stars (and the universe) would be very different. Because (1.9.29) is so slow, a catalytic process known as the CNO cycle proceeds more rapidly in stars more massive than the Sun, even though it requires reactions with $Z_1 Z_2 = 7$.

There are no stable nuclei with $A = 5$ or 8, so that helium nuclei cannot react with each other or with protons. More exotic reactions (such as ^{3}He + ^{4}He, or He + Li) also do not cross the $A = 8$ barrier. The only way to build nuclei heavier than $A = 8$ is by the process

$$\alpha + \alpha + \alpha \rightleftharpoons {}^{12}\mathrm{C}^* \rightarrow {}^{12}\mathrm{C} + \gamma + \gamma', \tag{1.9.30}$$

where the asterisk denotes the 7.654 MeV excited state and the right hand side indicates two successive radiative decays. This process is resonant because the energy of $^{12}C^*$ is only $E_* = 379$ KeV above that of three α-particles. In (1.9.30) the decay rate Γ_α of $^{12}C^*$ to the left is much faster than that Γ_γ to the right; the excited state is in thermal equilibrium with the α-particles, and its density n_* may be calculated from the Saha equation, with the result:

$$n_* = n_\alpha^3 \left(\frac{h^2}{2\pi k_B T}\right)^3 \left(\frac{3m_\alpha}{m_\alpha^3}\right)^{3/2} \exp(-E_*/k_B T), \qquad (1.9.31)$$

where n_α and m_α are the α-particle density mass.

The exponential in (1.9.31) contains the critical temperature dependence, which is characteristic of resonant reaction rates and is even steeper than that of (1.9.28). The factor P_0 need not be calculated explicitly because it enters in both directions on the left hand side of (1.9.30). A steady state abundance of $^{12}C^*$ is achieved in a time $\sim \Gamma_\alpha^{-1} \sim 10^{-15}$ sec. In practice, (1.9.30) proceeds through the unbound ^{8}Be nucleus (a scattering resonance only 92 KeV above the energy of 2 α-particles), rather than through a triple collision, but this does not affect the thermodynamic argument or the result. The reaction rate is $n_*\Gamma_\gamma$. The presence of an excited state of ^{12}C at the right energy to facilitate (1.9.30) is the reason carbon is a relatively abundant element in the universe; this is apparently fortuitous unless one attributes it to divine intervention, or argues that if it were not there we would not be present to observe its absence.

1.10 Polytropes

The solution of the equations (1.3.1–4) of stellar structure is complicated, because the equation of hydrostatic equilibrium (1.3.1) is coupled to the equation of energy flow (1.3.4) through (1.3.3) and the constitutive relation among P, ρ, and T. This problem is now readily handled numerically, even if some of the assumptions (most importantly, that of a thermal steady state) made in deriving (1.3.1–4) are relaxed. In the early (pre-computer) decades of stellar structure research this was not possible, and calculations of models simplified still further were performed. These methods are of more than historical interest, because the very simplified models which they produced are still powerful qualitative tools in understanding stars. They cannot replace modern computational methods of obtaining quantitative results, but they are much more transparent than a table of numbers, and therefore are very helpful to the astrophysicist who needs a qualitative understanding of the properties of self-gravitating configurations of matter.

A *polytrope* is a solution of the equation of hydrostatic equilibrium (1.3.1) under the assumption that the pressure P and the density ρ are everywhere related by the condition

$$P = K\rho^{\frac{n+1}{n}}. \tag{1.10.1}$$

The quantity n is called the polytropic index.

This relation is formally identical to the adiabatic relation (1.9.7) if $\gamma = \frac{n+1}{n}$, but their meanings are quite different. Equation (1.9.7) describes the variation of the properties of a fluid element undergoing an adiabatic process. Equation (1.10.1) constrains the variations of P and ρ with radius in a star, because if r is introduced as a parameter it relates $P(r)$ and $\rho(r)$. A star may be described by (1.10.1) even if the thermodynamic properties of its constituent matter are described by an adiabatic exponent γ different from $\frac{n+1}{n}$.

Equations (1.10.1) and (1.9.7) are equivalent if a star is neutrally stable (equivalently, marginally unstable) against convection, so that the actual dependence of P on ρ in the star is the same as the adiabatic one. This will be the case in a star which is completely convectively mixed, as is believed to be the case for very low mass main-sequence stars ($M \lesssim 0.2M_\odot$). The envelopes of red giants and supergiants are mixed, and also resemble polytropes if the gravitational influence of their dense cores may be neglected (a fair approximation if the envelope is very massive). In each of these cases $n \approx 3/2$; the deep convective envelope is a consequence of the high radiative opacity in the surface layers. Very luminous and massive stars also possess extensive mixed inner regions, and their envelopes are not far from convective instability. For these stars $n \approx 3$; convection is a consequence of their large luminosity.

The assumption of (1.10.1) in place of (1.3.4) permits the stellar structure equations to be reduced to a single nonlinear ordinary differential equation characterized by the parameter n. This equation is readily integrated numerically (even without computers!). Eliminating M from (1.3.1) and (1.3.2), we obtain

$$\frac{1}{r^2}\frac{d}{dr}\left(\frac{r^2}{\rho}\frac{dP}{dr}\right) = -4\pi G\rho. \tag{1.10.2}$$

Dimensionless variables are defined: $\phi^n \equiv \rho/\rho_c$ and $\xi \equiv r/\alpha$, where ρ_c is the central density, and the characteristic length (not the radius) $\alpha \equiv \left[\frac{(n+1)K\rho_c^{(1-n)/n}}{4\pi G}\right]^{1/2}$. Substitution of these variables and (1.10.1) into (1.10.2) yields the Lane-Emden equation:

$$\frac{1}{\xi^2}\frac{d}{d\xi}\left(\xi^2\frac{d\phi}{d\xi}\right) = -\phi^n. \tag{1.10.3}$$

The boundary conditions at $\xi = 0$ are $\phi = 1$ and $\frac{d\phi}{d\xi} = 0$. The surface is defined as the smallest value of ξ for which $\phi = 0$ (the solution for larger ξ is of no physical significance). Once a numerical integration in the dimensionless variables has been tabulated, it is readily applied to a star of specified ρ_c and K by using the definitions of ϕ and ξ.

Polytropes with certain values of n are of special interest. The ratios of the central density ρ_c to the mean density $\langle\rho\rangle$ indicate the degree to which mass is concentrated in their centers, and are a convenient one-parameter description of their structure.

If $n = 0$ then (1.10.1) corresponds to an incompressible fluid (only one value of ρ is permitted) and $\rho_c/\langle\rho\rangle = 1$. The definitions of ϕ, α, and K become indeterminate; with a little care they could be redefined, but there are easier ways of calculating the radius and pressure distribution of a sphere of incompressible fluid.

If $n = 1$ (1.10.3) is linear and may be integrated analytically, with the result $\phi = \sin\xi/\xi$. Here $\rho_c/\langle\rho\rangle = 3.29$.

If $n = 3/2$ (1.10.1) corresponds to an adiabatic star with $\gamma = 5/3$, and is therefore a good description of fully convective stars with this equation of state. The calculated $\rho_c/\langle\rho\rangle = 5.99$ is the lowest such value which may be obtained for stars composed of perfect gases.

If $n = 3$ (1.10.1) corresponds to an adiabatic star with $\gamma = 4/3$, and is therefore a good description of fully convective (or nearly convective) stars with this equation of state. It also turns out that an $n = 3$ polytrope is a fair description of the density structure $\rho(r)$ of stars in the middle and upper main sequence. Their deep interiors have steeper density gradients than they would if they were convective, but the adiabatic γ is larger than that of a fully convective $n = 3$ polytrope (for which γ must be 4/3); these two effects roughly cancel. For an $n = 3$ polytrope $\rho_c/\langle\rho\rangle = 54.2$. In the present-day Sun this ratio is calculated to be close to 100, while when the Sun was young it was about 60 (the difference results from the depletion of hydrogen and the increase in the molecular weight in the core). The structure and properties of an $n = 3$ polytrope are widely used when a rough but convenient model of a star is needed for more complex calculations.

If $n = 5$ (1.10.3) may also be solved analytically, with the result $\phi = (1 + \xi^2/3)^{-1/2}$. For $n \geq 5$ the radius is infinite because ξ never drops to zero.

If $n \to \infty$ (1.10.1) approaches an isothermal equation of state. The definition of ϕ becomes improper, but (1.10.2) is readily integrated without using (1.10.3). At large r, $\rho \propto r^{-2}$ and $M(r) \propto r$, so that both the radius and the total mass diverge. Such configurations do not describe stars. The upper atmospheres of stars may be isothermal but their structure does not approach an $n = \infty$ polytrope except at very large radii and extremely small density. Long before this the assumption of hydrostatic equilibrium will have failed because of the forces applied by the interstellar medium. These $n = \infty$ polytropes may describe the structure of gravitating clusters of collisionless objects (clusters of stars or of galaxies, for example).

1.11 Mass-Luminosity Relations

In **1.4** we derived scaling relations and made order-of-magnitude estimates for the characteristic ρ, P, and T of a star of given mass M and radius R. We now make similar approximations to estimate the relation between L and M of a main sequence star. As in **1.4**, our results are not meant to be numerically accurate, but rather to be an illuminating guide to the governing physics of stars of various masses.

We begin by defining β, the ratio of the gas pressure to the total pressure:

$$P_g = \beta P \tag{1.11.1a}$$

$$P_r = (1 - \beta)P. \tag{1.11.1b}$$

The parameter β is a function of T and ρ and, in general, varies from place to place within a star. Here we assume that it is a constant throughout a given star. This is true for an $n = 3$ perfect gas polytrope, because in such a polytrope the variations in ρ and T are related by $\rho \propto T^n$, so that the two terms in (1.4.5) vary in proportion. Stars on the middle and upper main sequence are approximately described as $n = 3$ polytropes, so that for them our results, derived assuming a constant β, are fair approximations to reality.

Now rewrite (1.3.4) or (1.7.19) in the form

$$\frac{dP_r(r)}{dr} = \frac{\kappa(r)\rho(r)L(r)}{4\pi c r^2}, \tag{1.11.2}$$

and divide this equation by (1.3.1). The result is

$$\frac{dP_r}{dP} = \frac{\kappa(r)L(r)}{4\pi cGM(r)}. \tag{1.11.3}$$

Drop the explicit dependence on r, and use (1.11.1) to rewrite this in terms of a constant β:

$$L = \frac{4\pi cGM}{\kappa}(1-\beta). \tag{1.11.4}$$

This equation is a fundamental relation among L, M, κ, and β. Because $\beta > 0$ it implies an upper limit on the radiative luminosity of a star.

In hot, luminous stars $\kappa \approx \kappa_{es}$ (1.9.17), so that

$$L = L_E(1-\beta), \tag{1.11.5}$$

where the Eddington limiting luminosity L_E is defined

$$\begin{aligned} L_E \equiv \frac{4\pi cGM}{\kappa_{es}} &= 1.47 \times 10^{38} \frac{\text{erg}}{\text{sec } M_\odot} \\ &= 3.77 \times 10^4 \left(\frac{M}{M_\odot}\right) L_\odot. \end{aligned} \tag{1.11.6}$$

Therefore, L_E is the upper limit to the radiative luminosity of hot stars. As discussed in **1.8**, it does not properly apply to the convective luminosity; it probably does still limit the luminosity of hot convective stars because their luminosity must flow radiatively through their atmospheres, where convection is ineffective. Cool supergiants may perhaps evade the limit (1.11.6) because κ may be very small in their cool atmospheres, but there is no evidence that they actually do so.

We can also express β in terms of ρ and T, and by so doing obtain a unique (though very approximate) relation between L and M. From the definitions of P_g, P_r, and P (1.4.5) we obtain, after eliminating T,

$$P = \left[\frac{3}{a}\left(\frac{N_A k_B}{\mu}\right)^4 \frac{1-\beta}{\beta^4}\right]^{1/3} \rho^{4/3}. \qquad (1.11.7)$$

Now use the relations (1.4.6,7) to express the dependence of P and ρ on M and R. In order to obtain a more useful numerical result we take the actual values of the coefficients which have been calculated for an $n = 3$ polytrope. The result is

$$\frac{1-\beta}{\beta^4} = 2.979 \times 10^{-3} \mu^4 \left(\frac{M}{M_\odot}\right)^2. \qquad (1.11.8)$$

This is known as Eddington's quartic equation. From it we may obtain $\beta(M)$ and $L(M)$. Note that β and L do not depend explicitly on R.

At low masses ($M\mu^2 \ll 20 M_\odot$, which includes nearly all stars) $\beta \to 1$ and $1-\beta \propto \mu^4 M^2$. From (1.11.4), dropping the μ dependence, we obtain the mass-luminosity relation for constant κ:

$$L \propto M^3; \qquad (1.11.9)$$

this describes main sequence stars with $\kappa \approx \kappa_{es}$ and holds for $M_\odot \ll M \ll 50 M_\odot$.

For stars of yet lower mass, κ is roughly described by Kramers' law (1.9.18). If we use (1.4.6,8) to determine T and ρ in Kramers' law, then

$$L \propto M^{11/2} R^{-1/2} \propto M^5, \tag{1.11.10}$$

where the last relation assumed $M \propto R$, which is implied by the approximation (**1.9.4**) that thermonuclear energy generation makes the central temperature nearly independent of M.

The Sun is very near the transition between (1.11.9) and (1.11.10), and has $\beta \approx 0.9996$. Very low mass stars ($M \lesssim 0.2 M_\odot$) are fully convective and their luminosity is determined by their surface boundary condition; the relations of this section do not apply.

Although these results are only approximate, it is evident that L is a steeply increasing function of M; massive stars are disproportionately luminous and short-lived, and low mass stars are disproportionately faint. Very massive stars are also much rarer in the Galaxy than low mass stars, so that they do not overwhelmingly dominate the total luminosity produced by stars; stars of moderate (Solar) mass are not insignificant. If one picks a photon of visible starlight in the Galaxy (or, similarly, chooses a star randomly on the sky), there is a significant chance that it will have come from a star of moderate mass. Very low mass stars, however, are so faint (1.11.10) that they contribute little to the starlight of the night sky.

For very large masses ($M \gtrsim 50 M_\odot$) $\beta \propto M^{-1/2} \rightarrow 0$ and $L \rightarrow L_E$, so that

$$L \propto M. \tag{1.11.11}$$

Stars this massive are very rare or nonexistent, but (1.11.11) represents a limiting relation which is approached by the most massive and luminous stars.

The relations in this section are inapplicable to stars far from the main sequence. In degenerate dwarfs the pressure is almost entirely that of electron degeneracy, which was not included in (1.4.5). As a result T is much lower than (1.4.8) would suggest for these dense stars, and L is lower by several orders of magnitude. This was a puzzle until electron degeneracy pressure was understood. White dwarfs slowly cool to a state in which $T = 0$, $\beta = 1$, and $L = 0$, in complete contradiction to (1.11.8).

The internal structures of giants and supergiants differ drastically from those of $n = 3$ polytropes, with $\rho_c/\langle\rho\rangle$ larger by many orders of magnitude. As a result, the approximate relations (1.4.6,7) fail completely. The structures of these stars are discussed in **1.13**. An analogue of (1.11.8) may be obtained if, instead of (1.4.7), we write

$$P \sim \frac{GM}{R_c}\frac{M}{R^3}, \tag{1.11.12}$$

where $R_c = \varsigma R$ is the core radius. Then we obtain

$$\frac{1-\beta}{\beta^4} \sim \frac{\mu^4 M^2}{\varsigma^3}. \tag{1.11.13}$$

Because $\varsigma \ll 1$, the limit $\beta \rightarrow 0$ is approached for much smaller M than would otherwise be that case; this crudely describes the high luminosity of giant and supergiant stars.

1.12 Degenerate Stars

The basic theory of cold degenerate stars was developed by Chandrasekhar, shortly after the development of quantum mechanics and the Pauli exclusion principle made possible the calculation of degenerate equations of state. His work was concerned with stars in which the electrons are degenerate, known to astronomers as white dwarves, and the discussion of this section generally refers to them. The results and conclusions are also qualitatively (but not quantitatively) applicable to neutron stars, in which degenerate neutrons contribute most of the pressure.

The theory of degenerate stars quantitatively predicts a relation between their masses and radii. It is possible to consider also a number of small effects not included in the basic theory, such as the effect of nonzero temperature, the structure of the nondegenerate atmosphere, the thermodynamics of the ion liquid and its crystallization, gravitational sedimentation in the atmosphere and in the deep interior, ... , and to make detailed predictions about luminosities, spectra, cooling histories, and other properties. Unfortunately, the quality of the extant data is inadequate to test either the basic mass-radius relation or these more sophisticated theories. Reliable masses are known for only a very few degenerate dwarves, and accurate radii for fewer (if any). Therefore, we are here concerned chiefly with their most basic properties, for which the theory, based only on quantum mechanics and Newtonian gravity, may be assumed with confidence.

In order to calculate the relation between the masses and radii of degenerate stars, we should calculate the zero-temperature equation of state $P(\rho)$ for arbitrary density, including the important regime of $\rho \sim 10^6$ gm/cm^3 lying between the relativistic (1.9.15) and nonrelativistic (1.9.14) limits. These calculations exist (see Chandrasekhar 1939), but a qualitative approach using the virial theorem may be more illuminating.

The total energy E of a star is

$$E = E_{grav} + E_{in}. \tag{1.12.1}$$

The quantitative value of each of these terms depends on the detailed forms of $\rho(r)$, $M(r)$, and $\mathcal{E}(r)$. Their scaling with M and R may be simply written, using relations like (1.4.6,7)

$$E_{grav} = -\int_0^R \rho(r)\frac{GM}{r}4\pi r^2\, dr \equiv -\mathcal{A}\frac{GM^2}{R} \tag{1.12.2a}$$

$$E_{in} = \int_0^R \mathcal{E}\, 4\pi r^2\, dr \equiv \mathcal{B}K\left(\frac{M}{R^3}\right)^\gamma R^3, \tag{1.12.2b}$$

where $\mathcal{A}$ and $\mathcal{B}$ are dimensionless numbers of order unity, and we have written $P = K\rho^\gamma \propto (M/R^3)^\gamma$, as is appropriate for adiabatic changes. For our qualitative considerations, we will assume that $\mathcal{A}$ and $\mathcal{B}$ are independent of changes in R, although this is not accurate except in the extreme nonrelativistic and extreme relativistic limits.

To compute the dynamical equilibrium radius of the star we find the minimum of the function $E(R)$. If $\gamma = 5/3$ there is a stable minimum E at

$$R = \frac{2\mathcal{B}K}{\mathcal{A}GM^{1/3}}. \tag{1.12.3}$$

This result is strictly applicable only in the limit $\rho \to 0$ (in order that $\gamma = 5/3$ hold exactly), $R \to \infty$, and $M \to 0$.

(1.12.3) describes the mass-radius relation of low mass degenerate dwarves, for which $\gamma = 5/3$ is a good approximation. (1.12.3) applies also to any series of $n = 3/2$ polytropes with a given value of K (equivalently, with a given entropy); if one adds to the outside of such a star matter with the same K as that inside, it will shrink. This is true both of degenerate dwarves (for which $S = 0$) and of low mass nondegenerate stars. The appearance of M in the denominator of (1.12.3) may be surprising; it is a consequence of the compressibility of matter and the increase of the gravitational force with increasing mass.

For small bodies, like those of everyday life, the density is set by their atomic properties, (1.9.14) is inapplicable, and $R \propto M^{1/3}$ (this may be taken as the definition of a planet). Jupiter is near the dividing line between these two regimes, and thus has approximately the largest radius possible for *any* cold body.

If $\gamma = 4/3$ the condition of minimum E is an equation for M, in which R does not appear:

$$M = \left(\frac{\mathcal{B}K}{\mathcal{A}G}\right)^{3/2}. \tag{1.12.4}$$

Such a configuration is an $n = 3$ polytrope, and $\mathcal{A}$ and $\mathcal{B}$ may be calculated from the known properties of polytropes. We know (see **1.5**) that if $\gamma = 4/3$ then $E = 0$, independently of R, so the absence of R from (1.12.4) is no surprise. Because the binding energy is zero and independent of R the radius is indeterminate.

More remarkable is the fact that a solution exists for only one allowable mass! This mass is called the Chandrasekhar mass M_{Ch}. Numerical evaluation for the relativistic degenerate equation of state (1.9.15) gives

$$\begin{aligned} M_{Ch} &= 5.75 M_\odot / \mu_e^2 \\ &\sim \left(\frac{\hbar c}{G m_P^2} \right)^{3/2} m_P. \end{aligned} \tag{1.12.5}$$

Calculations of stellar evolution and nucleosynthesis indicate that real degenerate dwarves will be composed principally of carbon and oxygen; in the special case in which they are built up by the gradual accretion of matter supplied from the outside they may be principally helium. For all of these elements the molecular weight per electron $\mu_e = 2$. M_{Ch} is reduced slightly below the value given in (1.12.5) by some small effects; the final numerical result is $M_{Ch} = 1.40 M_\odot$ (Hamada and Salpeter 1961).

The unique mass (1.12.4,5) and indeterminate radius apply only in the limit $R \rightarrow 0$ and $\rho \rightarrow \infty$, because only in this limit is $\gamma = 4/3$ exactly. Between this singular solution and the low density limit (1.12.3) there are solutions in which $4/3 < \gamma < 5/3$, and the equation of state is only partly relativistic. These solutions are not polytropes (because γ is not constant within them), but are readily calculated. Observed degenerate dwarves are believed to lie in the range $0.4 M_\odot \lesssim M \lesssim 1.2 M_\odot$, and to be in this semirelativistic regime. Calculations show that for these masses $R \approx 6000 (M_\odot / M)$ km is a fair approximation; their characteristic density is $\rho \sim 2 \times 10^6$ gm/cm^3 (1.9.16). By using the virial theorem (**1.5**) we can also estimate the surface gravitational potential $GM/R \sim m_e c^2$ (actual calculated values are $\sim$ 100 KeV/amu).

If $M > M_{Ch}$ no zero-temperature hydrostatic solutions exist. This is probably the most important result in astrophysics, because it means that stars more massive than M_{Ch} must either reduce their masses below M_{Ch}, end their lives in an explosion, or ultimately collapse.

Equations (1.12.3,4) apply to nondegenerate stars as well. For example, (1.12.4) describes the dependence of K on M for very massive stars, which approximate $n = 3$ polytropes because of the importance of radiation pressure. The factor K has larger values for nondegenerate matter than for degenerate matter, which has the lowest possible P at a given ρ.

The discussion of this section also applies qualitatively to neutron stars. Their characteristic density is determined by (1.9.16), and is $\sim (m_n/m_e)^3$ times larger than that of degenerate dwarves, and their radii are $\sim m_e/m_n$ times as large. Because K is independent of m_d in the relativistic regime (1.9.15), (1.12.4) predicts essentially the same limiting mass for neutron stars as for degenerate dwarves. Their surface gravitational potential $GM/R \sim m_n c^2$ (actual numerical values are believed to be $\sim$ 100 MeV/amu). The strong interactions between neutrons make (1.9.14,15) and (1.12.4) rough approximations at best; the equation of state of neutron matter is controversial. However, the conclusion that as $\rho \to \infty$ the Fermi momentum $p_F \to \infty$ and $\gamma \to 4/3$, which implies an upper mass limit M_{Ch}^{ns}, is inescapable. The effects of general relativity are also significant, and tend to increase the strength of gravity and to reduce M_{Ch}^{ns}, though they are not as large as the uncertainties in the equation of state.

Most calculations agree that for neutron stars $R \approx 10$ km, approximately independent of mass for $0.5 M_\odot \lesssim M \lesssim M_{Ch}^{ns}$. The value of M_{Ch}^{ns} is also controversial, but it is probably in the range

$1.40 M_\odot < M_{Ch}^{ns} \lesssim 2.5 M_\odot$. The lower bound on M_{Ch}^{ns} is firm, and is obtained from the observation of neutron stars of this mass in the binary pulsar PSR 1913+16, for which relativistic orbital effects permit accurate determination of the the pulsar mass (this is the only accurately determined neutron star mass). Because it is hard to imagine the production of neutron stars except as a consequence of the collapse of degenerate dwarves (or the degenerate dwarf cores of larger stars), it is likely that most neutron stars have $M \geq M_{Ch}$, which also implies $M_{Ch}^{ns} \geq M_{Ch}$. The upper bound on M_{Ch}^{ns} is less certain, but uncontroversial properties of the equation of state imply that it cannot much exceed $2.5 M_\odot$.

It is frequently pointed out in nontechnical astronomy books that a teaspoon (5 cm^3) of typical white dwarf matter has a mass of about 10 tons. It is not usually added that the internal energy of this teaspoonful is equivalent to that released by about 20 megatons of high explosive.

1.13 Giants and Supergiants

Main sequence and degenerate stars may be approximately described as polytropes. For giants and supergiants polytropic models and the rough approximations of **1.4** fail completely. These stars contain dense cores, resembling degenerate dwarves, and very dilute extended envelopes. The ratio $\rho_c/\langle\rho\rangle$, which is 54.2 for an $n = 3$ polytrope, may be $\sim 10^{12}$ (or more, in extreme cases).

The development of giant structure in a star is the outcome of complex couplings among the equations (1.3.1–7). Their solutions, obtained numerically, are the only proper explanation of giant structure, but it is useful to consider rough arguments. If the core and envelope are considered separately, the approximations of **1.4**, and simple models, may still be qualitatively informative.

A main sequence star will eventually exhaust the hydrogen at its center, leaving a core of nearly pure helium. For stars of masses approximately equal to or exceeding that of the Sun, this happens in less than the age of the Galaxy. Stars have presumably been born throughout that time (there are few quantitative data), so that there now exist stars of a variety of masses which have helium cores. Because the star continues to radiate energy, in a thermal steady state hydrogen must continue to be transformed to helium. This will happen in the hottest part of the star which contains hydrogen, a thin shell just outside the helium core.

The helium core will be essentially inert. In steady state it is isothermal at the temperature of the hydrogen burning shell at its outer surface. Because of the thermostatic properties (**1.9.4**) of thermonuclear energy release, we may roughly regard this shell as having a fixed $T_\circ \approx 4 \times 10^7$°K.

Once the core has accumulated a significant fraction (typically 8%) of the stellar mass, its temperature $T_\circ$ is insufficient to satisfy the equation (1.3.1) of hydrostatic equilibrium. Equation (1.4.8) explains why; T is set by the shell temperature, and hence by the structure of the outer star, but the core has a larger value of μ (4/3 for helium) and its higher density leads to a large M/R. It then contracts, producing a higher T (this process is stable, by the arguments of **1.5**). Now heat flows outward, which leads to yet higher T (the negative effective specific heat discussed in **1.5**). The heat flow reduces the entropy of the core, until its equation of state approaches that of a degenerate electron gas; the core comes to resemble a degenerate dwarf inside the larger star.

Core contraction will be interrupted when the temperature becomes high enough ($T \gtrsim 10^8$°K) for reaction (1.9.30) to take place, and exothermically to convert helium to carbon (auxiliary reactions also produce oxygen and rarer elements). This leaves an inert carbon-oxygen core surrounded by a double shell, the outer shell burning hydrogen and the inner shell burning helium. Such double shells have a complex and unstable evolution, but this is irrelevant to our rough description of the structure of a giant star.

The combination of a degenerate dwarf core with a thermostatic boundary condition produces the extended low density envelope of a giant star. A simple argument uses the scale height of the matter overlying the core. If L is not close to L_E radiation pressure is unimportant (see 1.11.5). An isothermal gas, supported in hydrostatic equilibrium by gas pressure against a uniform acceleration of gravity $g = GM_c/R_c^2$, has a density which varies as

$$\rho \propto \exp(-r/h), \tag{1.13.1}$$

where the scale height h is

$$\begin{aligned} h &= \frac{R_c^2 N_A k_B T}{GM_c \mu} \\ &= \frac{R_c k_B T}{E_b \mu}, \end{aligned} \tag{1.13.2}$$

and E_b is the gravitational binding energy per nucleon. The matter is not accurately isothermal and g is not strictly constant, but for $h \ll R_c$ these are good approximations. The approximations made in **1.4** were equivalent to assuming $h \sim R$ everywhere in the stellar interior, and fail at the core-envelope boundary where $h \ll R_c \ll R$.

For a degenerate core with $M_c = 0.7 M_\odot$, at $r = R_c$ we find $h/R_c \approx 0.055 \ll 1$. As a result, the density drops by a large factor in the region just outside the core boundary, where g is large. If the envelope contains a significant amount of mass, as it will in most giants, then this low density requires it to have a large volume and a large radius. Very crudely, we might expect the radius to be larger than that of a main sequence star (which the envelope would otherwise resemble) by a factor $\sim \exp\big(R_c/(3h)\big) \sim 10^2 - 10^3$, which is consistent with the radii of large red giants. If the core is more massive the density will be yet lower and the radius yet larger. The actual radius and T_e of a red giant are determined by the surface boundary conditions on its outer convective zone.

This argument is not applicable when $L \approx L_E$, because then the scale height is larger by a factor $\beta^{-1} \gg 1$. Instead, we equate the pressure of radiation to the pressure produced by the weight of the overlying matter, so that

$$\frac{a}{3}T_\circ^4 \sim \frac{GM_c\rho}{R_c}. \tag{1.13.3}$$

For $M_c = 1.2M_\odot$ ($\beta \ll 1$ only as $M_c \rightarrow M_{Ch}$) and $T_\circ = 4 \times 10^{7}$°K we estimate $\rho \sim 0.02$ gm/cm^3. If the envelope roughly resembles an $n = 3$ polytrope, as is likely, then its radius will be $\sim 20R_\odot$. Such a star is not nearly as large as a red giant or supergiant, but possesses a less extreme form of their structure of dense core and extended envelope. Because of its high luminosity and moderate radius its surface temperature is high. These stars are found in a region of the Hertzsprung-Russell diagram between the red supergiants and the upper main sequence, called the horizontal branch (most horizontal branch stars are produced differently, when rapid helium burning increases R_c and h).

1.14 Spectra

The study of astronomical spectra is a large field of research. Here we only draw a few qualitative conclusions useful in modelling novel objects and phenomena.

The radiation we observe from stars is produced in their atmospheres, and its spectrum reflects the physical conditions there. These atmospheres may usually be approximated as plane-parallel layers, so that in the equation (1.7.6) of radiative transfer we may neglect the term containing $1/r$. Then

$$\frac{\partial I(\tau,\vartheta)}{\partial \tau}\cos\vartheta - I(\tau,\vartheta) + S(\tau) = 0, \tag{1.14.1}$$

where the source function $S(\tau) \equiv j(\tau)/4\pi\kappa(\tau)$, and the optical depth τ is defined by $d\tau \equiv \kappa\rho dr$, and $\tau \to 0$ as $r \to \infty$. I, j, κ, and τ all implicitly depend on ν. For $\cos\vartheta > 0$ this equation has the formal solution

$$I(\tau,\vartheta) = \int_\tau^\infty S(\tau')\exp\left[-(\tau'-\tau)\sec\vartheta\right]\sec\vartheta\, d\tau'. \tag{1.14.2}$$

The emergent flux is that at $\tau = 0$:

$$I(0,\vartheta) = \int_0^\infty S(\tau')\exp(-\tau'\sec\vartheta)\sec\vartheta\, d\tau'. \tag{1.14.3}$$

The emergent flux is a weighted average of S over the atmosphere, with most of the contribution coming from the range $0 \leq \tau' \lesssim \cos\vartheta$.

The opacity κ_ν of matter typically has the form shown in Figure 1.3, with sharp atomic lines superposed on a slowly varying continuum. The lines are those of the species abundant in the atmosphere, which depend on its chemical composition, density, and (most sensitively) temperature. In hot stars the strong lines are those of species like He II and C III, in somewhat cooler stars those of He I or H I, in yet cooler stars Ca I and Fe I, and in the coolest stars those of molecules like TiO.

In the simplest stellar atmospheres matter is in thermodynamic equilibrium, there is no scattering, $S = B$ (the Planck function), and the temperature increases monotonically inward. Then I reflects the value of B in the region $\tau \sim 1$, and we may approximate $I_\nu(\tau = 0) \approx B_\nu(T(\tau_\nu = 2/3))$. At a line frequency ν_l the opacity κ_{ν_l} is large and $\tau_{\nu_l} = 2/3$ high in the atmosphere, where T and B are low,

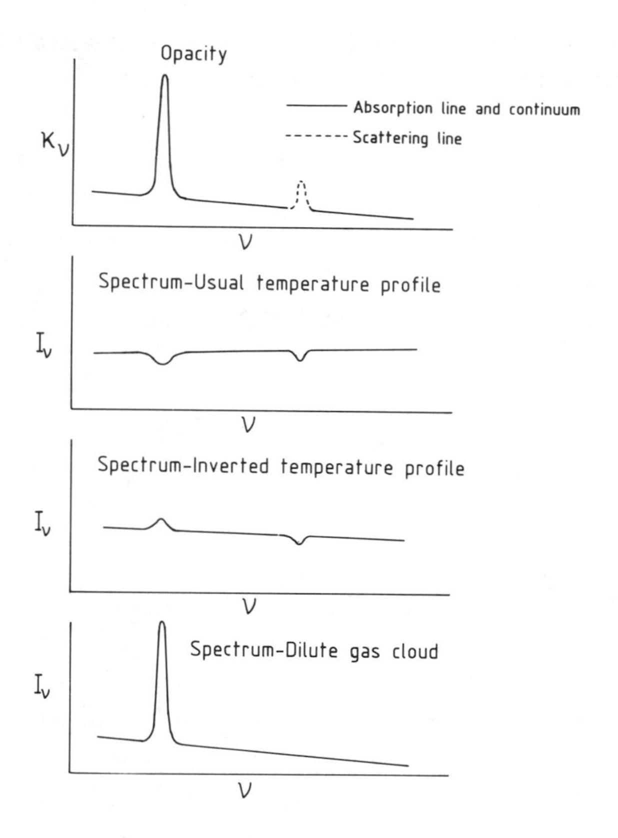

Figure 1.3. Varieties of Spectra.

while outside the line κ_ν is small and $\tau_\nu = 2/3$ much deeper in the atmosphere. The result is an absorption line spectrum, as shown in the figure.

In many stars the upper atmosphere is much hotter than the rest of the atmosphere. In the Sun the upper atmosphere and corona are heated by acoustic (or magneto-acoustic) waves generated within the convective zone. In a few stars a strong radiation flux from a luminous binary companion heats the upper atmosphere; this is found in some companions to strong X-ray sources. When the temperature profile is inverted in this manner there results an emission line spectrum, as shown in the figure. Often a weak emission line spectrum from the highest levels of the atmosphere is superposed on a stronger absorption line spectrum.

If line scattering opacity is important it may also produce an absorption line, regardless of the temperature gradient in the atmosphere. The mechanism is outlined in **1.7.3**; the presence of scattering reduces the emissivity. At such frequencies the diffuse reflectivity of the atmosphere is significant, so that a fraction of the flux is the (zero) reflected flux of the dark sky. If there is significant scattering opacity in the continuum, but the line opacity is absorptive, then the sky is reflected in the continuum and the line will appear in emission. These processes are known as the Schuster mechanism.

In a dilute gas cloud the upper limit in the integral (1.14.3) is τ_{max}, the total optical depth integrated through the cloud. Often the cloud is so rarefied and transparent that $\tau_{max} \ll 1$ at all frequencies. Then (1.14.3) may be approximated

$$I(0,\vartheta) \approx \frac{j_\nu}{4\pi} \sec\vartheta \int \rho \, dr. \qquad (1.14.4)$$

The frequency dependence of the emergent spectrum is that of the emissivity j_ν. Under these conditions LTE is usually inaccurate; the emergent spectrum qualitatively resembles that of the opacity κ_ν, although quantitative results require a calculation of the various atomic and ionic processes. There is an emission line spectrum in which the lines are extremely strong, carrying a significant fraction of the total flux. Such spectra are observed from interstellar clouds, winds flowing outward from stars, the debris of stellar explosions, stellar coronae, laboratory gas discharge lamps, and in other circumstances in which $\int \rho \, dr$ is very small. Because the emitting volume may be large, the total mass and radiated power need not be small, despite the low density.

These classes of spectra are very different, and may often be identified at a glance, even though they are not usually found in their pure states. This is useful in attempting to construct a rough model of a novel astronomical object, because the densities, dimensions, and directions of energy flow are readily constrained. Images are not available for many interesting astronomical objects, because of their small angular sizes, so that the first step in understanding them is the identification of their components and the construction of a rough model of their geometry, their physical parameters, and of the important physical processes.

1.15 Mass Loss

Spectroscopic observations show that many stars lose mass. Typically, the observations show emission lines whose Doppler widths indicate the flow velocity. In most cases the line shape does not directly establish that the mass is flowing outward, only that the star is surrounded by a dilute cloud of gas with the appropriate distribution of velocities; it is usually not possible to determine from the data which velocities are found at which points in space, but outflow is often the only plausible interpretation. In some cases the outflowing gas absorbs an observable amount of the stellar line radiation, and the resulting complex (P Cygni) line profiles may be interpreted unambiguously as mass outflow.

Some stars are observed in ordinary photographs (or infrared images) to be surrounded by luminous gas clouds they have expelled; in some cases these clouds have visibly expanded since the first photographs were taken. Many different kinds of stars lose mass by a variety of mechanisms and at widely varying rates. Even the Sun loses mass at the very small rate of $\sim 10^{-15} M_\odot$/year in the Solar wind, produced by the thermal expansion of its hot corona. All stars with convective surface layers are expected to have coronae, whose mass loss rates should be much greater in larger stars with lower surface gravity.

It is known that some stars born with M substantially larger than M_{Ch} have evolved into degenerate dwarves; this establishes that, in some cases, a star may lose the greater part of its mass. In this section I briefly and qualitatively discuss mass loss mechanisms which may occur in luminous stars, where the mass loss rate is often high. Most of these processes are not understood quantitatively.

In a very luminous star the radiation pressure approaches the total pressure, and $\beta \rightarrow 0$ (1.11.1). How closely a star approaches the neutrally stable limit $\beta = 0$ depends on the detailed calculation of its structure; we know (see 1.11.8) that very massive stars and giant stars with dense degenerate cores have small β. From the equation of hydrostatic equilibrium we have

$$-\frac{\beta G M \rho}{r^2} = \frac{dP_g}{dr}, \tag{1.15.1}$$

so that in this limit the gradient of the gas pressure becomes zero. Essentially the entire weight of the matter is supported by the gradient of radiation pressure; in other words, the force of gravity and the force of radiation pressure cancel. If $\beta = 0$ exactly, nothing is left to resist the gradient of P_g, and the stellar material will float off into space. This argument suggests that very luminous stars are likely to lose mass.

This conclusion is at least qualitatively correct, and may be reached on simple energetic grounds by noting that as $\beta \rightarrow 0$ we have $\gamma \rightarrow 4/3$, and that if $\gamma = 4/3$ the binding energy $E = 0$ (see **1.5**). It is possible to show, by manipulation of the stellar structure equations, that $n = 3$ polytropes (which stars approach as $\beta \rightarrow 0$)

with a constant β are neutrally stable against convection if $\gamma = 4/3$ (also approached as $\beta \rightarrow 0$); it is unsurprising that a star with zero binding energy should be neutrally stable against the interchange of its parts.

Should L exceed $4\pi cGM/\kappa$, the star becomes unstable against convection, and if convection is efficient it carries the excess flux. The radiative luminosity does not exceed $4\pi cGM/\kappa$ and the gradient of radiation pressure does not exceed the force of gravity. In fact, $L > 4\pi cGM/\kappa$ in the envelopes of many cool giants and supergiants, where κ is large; these stars generally do not lose mass rapidly. Only if convection is incapable of carrying the heat flux does excess radiation pressure drive a mass efflux.

It is comparatively easy to disrupt a star with $\beta \ll 1$ if it can be disturbed, but reliable calculation is difficult. Possible disturbances include fluctuations and instability in the nuclear energy generation rate (known to occur in supergiants with degenerate cores and double burning shells), and the inefficient convection present in the outer layers of cool giants and supergiants. Such stars may lose their entire envelopes in response to modest disturbances (most notably in the formation of planetary nebulae by supergiant stars), but it is also necessary to consider less dramatic mass loss processes. These are easier to observe (because they last longer) and to calculate.

The most important factor leading to steady mass loss is probably an increase in κ in optically thin regions above the photosphere. Because the density and optical depth are low, convection cannot transport heat effectively, and probably does not take place. Instead, matter can actually be subject to a force of radiation pressure exceeding that of gravity (a situation which would not occur in a stellar interior in hydrostatic equilibrium). At least two kinds of physical processes, changing ionization balance and grain formation, may produce such an abrupt jump in κ.

The temperature of a grey body (one whose opacity is independent of frequency) just outside a photosphere will be lower than that of one just inside by a factor of about $2^{-1/4} = 0.84$; outside, the black body radiation field only fills the 2π steradians of outward-directed rays, while the 2π steradians of inward-directed rays have little intensity. The opacity of stellar atmospheres is not accurately grey, but this is still a reasonable estimate of the temperature drop. Such a drop may be sufficient to substantially shift the ionization balance, and therefore the opacity. In addition, the Rosseland mean opacity, derived for stellar interiors (in which $\tau_\nu \gtrsim 1$ at all frequencies) is inapplicable in optically thin regions. In the opposite limit, $\tau_\nu \lesssim 1$ at all frequencies, the radiation force is proportional to $\int I_\nu \kappa_\nu \, d\nu$; the arithmetic mean opacity exceeds the Rosseland mean. Strong atomic or ionic lines may now make a large contribution to the force of radiation pressure, and calculations show that in the upper atmospheres of hot luminous stars the net acceleration may be upward.

A simple argument makes it possible to estimate the mass efflux. Suppose the matter is accelerated by radiation pressure in a spectral line of rest frequency $\nu_\circ$. Radiation between $\nu_\circ$ and $\nu_\circ(1 - v/c)$ may be absorbed or scattered by the outflowing wind; the total pressure the radiation field can exert on the matter may be $\sim H_{\nu_\circ} \nu_\circ v/c^2$. Equate this to the momentum efflux rate per unit area $\dot{m}v$ (where $\dot{m}$ is the rate of mass loss per unit area) to obtain the total mass loss rate $\dot{M}$:

$$\begin{aligned} \dot{M} &= 4\pi R^2 \dot{m} \\ &\sim 4\pi R^2 H_{\nu_\circ} \nu_\circ / c^2 \\ &\sim L/c^2, \end{aligned} \tag{1.15.2}$$

where we have approximated $L \equiv \int H_\nu \, d\nu \approx H_{\nu_\circ} \nu_\circ$. This result is an upper bound, because not all of the radiation at the frequencies of the Doppler-shifted line will be absorbed or scattered, and because gravity has been neglected. $\dot{M}$ is independent of v; calculations usually show that v is a few times the stellar surface escape velocity.

If N strong lines contribute to the absorption of radiation, then $\dot{M}$ may be larger by a factor N, which may be $\gg 1$, but not by orders of magnitude.

The mass efflux rate (1.15.2) is small, although readily observable spectroscopically. It is roughly the same as the equivalent mass carried off by the radiation field itself; we know that during a star's life thermonuclear reactions convert less than 1% of its mass to energy. If $L \approx L_E$ then $\dot{M} \lesssim 10^{-9} M/\text{year}$.

A luminous star with a very cool surface (a red supergiant) may lose mass in a related, but more effective way. Above its photosphere the temperature may be cool enough for carbon (and other elements or molecules) to condense into grains; this is probably the origin of interstellar grains. These grains (in particular, those of carbon) are very effective absorbers of visible and near-infrared radiation across the entire spectrum ($\kappa \sim 10^5$ cm^2/gm), so that the pressure of the radiation on the matter may be $\sim \int H_\nu \, d\nu / c$; the Doppler shift factor $v/c \ll 1$ does not enter. Then we obtain

$$\dot{M} \sim \frac{L}{vc}. \tag{1.15.3}$$

For a red supergiant $v \sim \sqrt{GM/R} \sim 30$ km/sec, so this result is $\sim 10^4$ times as large as (1.15.2). The time required to halve M may be as short as $\sim$ 30,000 years. Such a large mass loss rate may change the evolutionary history of the star; for example, it may reduce M below M_{Ch}. Unfortunately, it has not been possible to quantitatively calculate mass loss by this process, although observations indicate it does take place.

The highest estimate of mass loss comes if the energy of the star's radiation may be efficiently used to overcome the gravitational binding energy and to provide kinetic energy, so that

$$\dot{M} \sim \frac{L}{v^2}. \tag{1.15.4}$$

In order for this to occur the radiation must be trapped between an expanding optically thick outflow and the luminous stellar core, and be the working fluid in a heat engine. The required optical depth at all frequencies is $\tau \gtrsim c/v \gg 1$; the acceleration occurs in the stellar interior rather than in the atmosphere. However, such a radiatively accelerated optically thick shell will probably be unstable to convection if it is in hydrostatic equilibrium. Mass loss rates as high as (1.15.4) may be obtained when hydrostatic equilibrium does not apply; for example if L rises significantly above L_E in a time $< t_h$. Such an event resembles an explosion rather than steady mass loss.

Rapid astronomical processes are hard to observe directly, because the fraction of objects undergoing then at any time is inversely proportional to their duration. There is much less direct evidence for mass loss at the rates of (1.15.3) or (1.15.4) than at the slow rate (1.15.2), but the more rapid processes may be important in many objects; the formation of planetary nebulae is a probable example.

1.16 References

Bahcall, J. N., Huebner, W. F., Lubow, S. H., Parker, P. D., and Ulrich, R. K. 1982, *Rev. Mod. Phys.* **54**, 767.

Chandrasekhar, S. 1939, *An Introduction to the Study of Stellar Structure* (Chicago: University of Chicago Press).

Clayton, D. D. 1968, *Principles of Stellar Evolution and Nucleosynthesis* (New York: McGraw-Hill).

Hamada, T., and Salpeter, E. E. 1961, *Ap. J.* **134**, 683.

Harris, M. J., Fowler, W. A., Caughlan, G. R., and Zimmerman, B. A. 1983, *Ann. Rev. Astron. Ap.* **21**, 165.

Mihalas, D. 1978 *Stellar Atmospheres* 2nd ed. (San Francisco: W. H. Freeman).

Rees, M. J. 1966, *Nature* **211**, 468.

Salpeter, E. E. 1964, *Ap. J.* **140**, 796.

Schwarzschild, M. 1958, *Structure and Evolution of the Stars* (Princeton: Princeton University Press).

Chapter 2

Non-Equilibrium Thermodynamics

2.1 Kinetic Equations

A great many astrophysical problems involve distributions of particles which are not in thermodynamic equilibrium. In one sense this statement is trivial and tautological—the entire universe would be in equilibrium only if "heat death" had occurred. That clearly is not the case, and the universe would not be very interesting if it were, nor would we be around to care. More seriously, what is meant is that the local distributions of particles are frequently not close to equilibrium, and that the ways in which they approach equilibrium are interesting. This is not often the case in everyday life, where matter densities and collision rates are high, so that relaxation to equilibrium is rapid and deviations from it are small (a striking everyday exception is radiation; because the atmosphere is transparent, the sky is dark, and the Sun is bright, the visible radiation field around us is very far from any equilibrium distribution). In contrast, many astrophysical problems involve very low densities and low collisional relaxation rates, and deviations from equilibrium are large, important, and interesting (stellar interiors are an exception). Matter which is not in thermodynamic equilibrium is described by kinetic equations; a standard textbook is that by Liboff (1969).

In order to describe completely the state of a classical gas of identical particles which are not in thermodynamic equilibrium it is necessary to specify the coordinates and momenta of all its particles. The probability that one particle will be found in a cell of volume d^3x_1 at position $\vec{x}_1$, and in a momentum cell of volume d^3p_1 at momentum $\vec{p}_1$, that a second particle will be in a volume d^3x_2 at $\vec{x}_2$, and d^3p_2 at $\vec{p}_2$, ..., and so on, up to the N-th, is given by an N-particle distribution function f_N multiplied by the volume in phase space:

$$f_N(\vec{x}_1, \ldots, \vec{x}_N, \vec{p}_1, \ldots, \vec{p}_N, t)\, d^3x_1 \ldots d^3x_N d^3p_1 \ldots d^3p_N. \tag{2.1.1}$$

Clearly f_N is a horribly complex and unwieldy object, for N is typically of the order of Avogadro's number.

If the motion of any one particle is completely independent of any and all other particles, then instead of specifying f_N it is sufficient to specify a single particle distribution function f, where

$$f(\vec{x}, \vec{p}, t)\, d^3x d^3p \tag{2.1.2}$$

is the probability of finding a single particle in the phase space volume d^3xd^3p around the point in phase space $(\vec{x}, \vec{p})$; equivalently, f is the mean or expected density of particles there. The radical approximation of using (2.1.2) instead of (2.1.1) is exact for noninteracting particles and good for weakly interacting particles. It is usually justified for dilute gases, but is quite wrong for liquids and solids.

The single particle distribution function (2.1.2) is simple enough to be tractable. If there are no interactions between the particles and they are nonrelativistic it obeys the Liouville equation

$$\frac{df}{dt} = \frac{\partial f}{\partial t} + \frac{\vec{p}}{m} \cdot \frac{\partial f}{\partial \vec{x}} + \vec{F}(\vec{x}, \vec{p}, t) \cdot \frac{\partial f}{\partial \vec{p}} = 0, \tag{2.1.3}$$

where $\vec{F}$ is any force field which may be present, and the arguments $(\vec{x}, \vec{p}, t)$ of f have not been written for the sake of brevity. This equation is only the statement that because particles are conserved so is the particle density in phase space.

There are two ways in which (2.1.3) must be modified to include interactions between particles, because these interactions may be divided into two kinds. The first kind is the interaction of a single particle with the mean distribution of the other particles. If we assume that the mean distribution is completely specified by (2.1.2) then its effect on f may be included by adding a force field which depends on f, so that (2.1.3) now takes the form

$$\frac{df}{dt} = \frac{\partial f}{\partial t} + \frac{\vec{p}}{m} \cdot \frac{\partial f}{\partial \vec{x}} + \vec{F}(\vec{x}, \vec{p}, t, f) \cdot \frac{\partial f}{\partial \vec{p}} = 0. \tag{2.1.4}$$

This is known as the Vlasov equation, and is widely used in both plasma physics and stellar dynamics. F now includes the electric, magnetic and gravitational fields produced by the mean distribution of particles f, and is calculated from additional equations (Newton's or Maxwell's).

The second kind of interparticle interaction has a random and statistical character, and so cannot be described by a mean force field $\vec{F}$. Instead, it is assumed to take the form of instantaneous collisions between two particles, independent of any other particles in the fluid. This assumption is again exact in the trivial case of non-interacting particles, and is a good approximation for dilute gases with short range forces. It is generally a reasonable approximation in plasmas where the interaction is long range but screened at long distances; this screening, properly a three-particle interaction, is reasonably accounted for by modifying the interparticle force law. The assumption of instantaneous binary collisions is also believed to be a good approximation for systems of gravitating particles if a large number of particles are present.

Collisions remove particles from a cell in phase space (or supply them to it), so in (2.1.3) and (2.1.4) they take the form of a source or sink term in place of the zero on the right hand side. We first calculate the rate at which collisions scatter particles out of the phase space volume d^3xd^3p about the point $(\vec{x}, \vec{p})$.

Consider a collision between particles with momenta $\vec{p}$ and $\vec{p}_1$, producing particles with momenta $\vec{p}\,'$ and $\vec{p}_1'$. Ignore quantum statistics, so that the particles are distinguishable, though of the same kind. In order to have a collision with impact parameter between s and ds in a time dt the colliding particle must be found in a cylindrical volume element $2\pi s ds dt|\vec{p} - \vec{p}_1|/m$; before collision the first particle is on the axis of this cylinder, which is oriented along the relative velocity vector $(\vec{p} - \vec{p}_1)/m$. The rate at which collisions remove particles of momentum $\vec{p}$ is then

$$\int f_2(\vec{x}, \vec{x}, \vec{p}, \vec{p}_1)\; d^3p_1\; 2\pi s\; ds \left|\frac{\vec{p} - \vec{p}_1}{m}\right|, \tag{2.1.5}$$

where $f_2(\vec{x}, \vec{x}, \vec{p}, \vec{p}_1)$ is the two particle distribution function, giving the probability that there are two particles simultaneously present in a unit volume of phase space at coordinate $\vec{x}$ and at momenta $\vec{p}$ and $\vec{p}_1$.

In order to make any progress with (2.1.5) it is necessary to have (or at least assume) some information about f_2. It is possible to write an equation like (2.1.3) for f_2, and then to add a collision term to its right hand side. This would be useless, because the collision term would involve the rate at which collisions with a third particle change f_2; the rate of such triple collisions depends on f_3. An infinite hierarchy of equations would result, unless drastic action were taken. This resembles the closure problem of the moment equations of radiative transport theory (**1.7**), but is much worse, because of the increasing complexity of successive f_n.

Boltzmann took drastic action. He approximated

$$f_2(\vec{x}, \vec{x}, \vec{p}, \vec{p}_1) = f(\vec{x}, \vec{p}) f(\vec{x}, \vec{p}_1). \tag{2.1.6}$$

This is known as the "assumption of molecular chaos," and is justified in the dilute gas approximation we have been using all along. It says that there is no correlation between the momentum distributions of two colliding particles. It is not possible to justify this rigorously, but it is believed to hold in a dilute fluid because between scatterings a particle travels many times the mean nearest neighbor distance. The scattering particle (incident with momentum $\vec{p}_1$) almost certainly has scattered from many other particles and has thoroughly randomized its velocity since the previous time (if ever) it scattered from the particle of momentum $\vec{p}$ whose distribution we are calculating. The scattering particle may therefore be considered to have been drawn randomly from the particle distribution function f, so that its momentum is uncorrelated with that of the test particle from which it scatters. In contrast, in a liquid where two particles may find themselves encaged by their neighbors and repeatedly scattering from each other, and the assumption of molecular chaos would not be justified.

The total rate of removal of particles from the phase space cell $d^3x d^3p$ is

$$\int f(\vec{x}, \vec{p}) \, d^3p_1 \; 2\pi s \, ds \left| \frac{\vec{p} - \vec{p}_1}{m} \right|. \tag{2.1.7}$$

From this we must subtract the rate of the inverse process of scattering from momenta $\vec{p}\,'$, $\vec{p}_1'$ to $\vec{p}$, $\vec{p}_1$, which is clearly

$$\int f(\vec{x}, \vec{p}\,') f(\vec{x}, \vec{p}_1') \; 2\pi s' \, ds' \left| \frac{\vec{p}\,' - \vec{p}_1'}{m} \right|. \tag{2.1.8}$$

(2.1.8) is the rate of removal from the phase space cell $d^3x' d^3p'$. For a nonrelativistic elastic collision $|\vec{p} - \vec{p}_1| = |\vec{p}\,' - \vec{p}_1'|$, $s \, ds = s' \, ds'$,

$d^3p_1 = d^3p_1'$, $d^3p = d^3p'$, and $d^3x = d^3x'$. Then (2.1.7) and (2.1.8) may be combined and added to (2.1.3), dropping the explicit $\vec{x}$ dependence, to give the Boltzmann equation

$$\frac{df}{dt} = \frac{\partial f}{\partial t} + \frac{\vec{p}}{m} \cdot \frac{\partial f}{\partial \vec{x}} + \vec{F} \cdot \frac{\partial f}{\partial \vec{p}}$$
$$= \int d^3p_1 \left| \frac{\vec{p} - \vec{p}_1}{m} \right| \frac{d\sigma}{d\Omega} \, d\Omega \, \left[f(\vec{p}_1')f(\vec{p}') - f(\vec{p}_1)f(\vec{p}) \right], \tag{2.1.9}$$

where $\vec{p}_1'$ and $\vec{p}'$ are implicitly functions of $\vec{p}_1$, $\vec{p}$, and the scattering angle Ω (determined by the conservation of momentum and energy), and $d\sigma \equiv 2\pi s \, ds$. As before, in most terms the arguments of f are not written explicitly unless they differ from $(\vec{x}, \vec{p}, t)$. There is an implicit relation between s and Ω, which may be calculated from the force law between the particles; the differential cross-section $d\sigma/d\Omega$ may be obtained from this relation. Because we have ignored quantum mechanics, scattering is deterministic; (2.1.9) is also correct for quantum mechanical $d\sigma/d\Omega$, but our elementary derivation involving impact parameters is inapplicable.

The Boltzmann equation (2.1.9) is a nonlinear integrodifferential equation, and is therefore rather difficult to solve. It is harder to solve than the equation of radiation transport for scattering opacity, because the scattering of radiation by matter is linear in the radiation field (see 1.7.14), while the right hand side of (2.1.9) is nonlinear in f. The difficulty lies partly in the nonlinearity and partly in the fact that $df(\vec{p})/dt$ is related to an integral of $f(\vec{p}_1)$ over all values of $\vec{p}_1$; equivalently, one may say that the scattering is nonlocal in momentum space. If f is known a brute-force numerical evaluation of the right hand side is not difficult, but this is usually not very illuminating.

Certain special cases may be solved explicitly, at least to a useful approximation. In one such case f is everywhere nearly Maxwellian, but the mean fluid parameters of density, velocity, and temperature are slowly varying functions of space. Then an expansion (by Chapman and Enskog; see Chapman and Cowling 1960) of f about a Maxwellian leads to a linear integral equation which is easier to solve, and eventually to the calculation of the macroscopic transport coefficients. For a gas containing only one species of uncharged particle these are the thermal conductivity and the viscosity. If more than one species is present there are also diffusion coefficients and additional coefficients (thermal diffusion, diffusion thermo-effect, and cross-diffusion) coupling the fluxes of the several scalar quantities (heat and the concentrations of the species). If the particles are charged these are also coupled to the flow of current and the electrostatic potential.

This procedure is a remarkable accomplishment of formal nonequilibrium statistical mechanics, but it is often inadequate for the astrophysicist. In many astrophysical problems the dimensions are so large and gradients of temperature, velocity, and composition are so small that the flow of heat, momentum, and material by these processes is insignificant. Notable exceptions include heat conduction in white dwarfs and neutron stars and diffusion in some stellar photospheres. In many other problems the transport of heat, momentum, and matter are important, but the density is usually so low and the mean free paths to collisions so long that the distribution functions are very far from Maxwellian, and the Chapman-Enskog procedure (and the textbook values of transport coefficients) are inapplicable. One example of this is heat conduction in the Solar wind, in which electron mean free paths may be longer than the size of the solar system. Processes other than two body collisions, such as the excitation of plasma waves, then become dominant in determining the actual flow of heat. Another example is fluid turbulence. Locally the particle distribution functions are nearly Maxwellian, but on larger spatial scales the fluid velocity, temperature, (and sometimes composition) vary irregularly in both space and time. The result is a turbulent flow which transports heat, momentum, (and sometimes composition) many orders of magnitude faster than the microscopic transport processes. Unfortunately, no quantitative understanding of these turbulent transport processes exists. Semi-empirical models (such as "mixing-length theory" **1.8**) work in the laboratory and are very useful in engineering, where they can be tested and calibrated by experiment. Their extrapolation to flows on astrophysical scales, though widely assumed, is a matter of guesswork.

A second special case of the Boltzmann equation is very useful in astrophysics. This is the limit in which $\vec{p}'$ is close to $\vec{p}$, and $\vec{p}_1'$ is close to $\vec{p}_1$; collisions individually produce small momentum transfers. If the potential between particles is $\propto 1/r$, as is the case for electrostatic and gravitational interactions, then the integral in (2.1.9) is dominated by distant collisions which individually produce small momentum transfers. It is then possible to expand the right hand side of (2.1.9) in powers of $\Delta\vec{p} = \vec{p}' - \vec{p}$, and to carry out the integral. The result involves derivatives of f which enter from the Taylor series expansion of f about $f(\vec{p})$, multiplied by algebraic functions of $\vec{p}$ and f. The resulting differential equation is much easier to deal with. The calculation of charged particle equilibration in **2.2** and the derivation of the Kompaneets equation in **2.3** are examples of this kind of procedure. The result of an expansion of (2.1.9) in powers of small momentum transfers, including only first and second derivatives, is of the form:

$$\begin{aligned} \frac{df}{dt} &= \frac{\partial f}{\partial t} + \frac{\vec{p}}{m} \cdot \frac{\partial f}{\partial \vec{x}} + \vec{F} \cdot \frac{\partial f}{\partial \vec{p}} \\ &= -\frac{\partial}{\partial \vec{p}} \cdot \Big(\vec{a}(\vec{p}) f \Big) + \frac{1}{2} \frac{\partial^2}{\partial \vec{p} \partial \vec{p}} : \Big(\mathrm{b}(\vec{p}) f \Big), \end{aligned} \tag{2.1.10}$$

where the double dot product is between a tensor and a tensor operator. It is necessary to include second derivatives in the expansion of f, but generally not higher derivatives.

Equation (2.1.10) is purely formal; all the physics is contained in the coefficients $\vec{a}(\vec{p})$ and $\mathbf{b}(\vec{p})$, and their calculation is usually quite difficult (the derivation of the Kompaneets equation is one of the easier examples). In general, they will depend on f. Equations of the form of (2.1.10) are known as Fokker-Planck equations. They are powerful tools in the study of nonequilibrium distribution functions, and are generally valid whenever the evolution of a particle's momentum (or other conserved dynamical parameters) is the result of many small independent random events.

The significance of the coefficients $\vec{a}$ and $\mathbf{b}$ is apparent. Consider only the $\vec{a}$ term for a spatially homogeneous f with no external force. Then

$$\frac{\partial f}{\partial t} + \frac{\partial}{\partial \vec{p}} \cdot (\vec{a} f) = 0. \tag{2.1.11}$$

This is the equation of conservation of particles, if $\vec{a}$ is their "velocity" in momentum space (the rate at which their momentum is changing, or the mean force on them resulting from collisions). Therefore, $\vec{a}$ is generally called the coefficient of dynamical friction. It is usually directed opposite to $\vec{p}$, as a particle loses its initial momentum in randomizing collisions.

If we consider only the $\mathbf{b}$ term in (2.1.10), again for a spatially homogeneous f with no external force, and also take $\mathbf{b}$ to be a scalar constant b times the unit tensor, we obtain

$$\frac{\partial f}{\partial t} = \frac{1}{2} b \nabla_p^2 f, \tag{2.1.12}$$

where ∇_p^2 is the Laplacian with respect to the momentum coordinates. We see that b represents a diffusion coefficient in momentum space, as a particle suffers a random series of impulses. If $f(\vec{p})$ is very narrowly peaked, then its second derivatives are much larger than its first derivatives, and the peak spreads out in a Gaussian shape according to (2.1.12). As the peak broadens the $\vec{a}$ term becomes significant, and begins to counteract the broadening.

If f is evolving by collisions with a background equilibrium distribution of particles at temperature T, then a stationary solution to (2.1.10) must be proportional to a Maxwellian distribution $f_0 \equiv \exp(-p^2/2mk_BT)$ at that temperature. The coefficients $\vec{a}$ and $\mathbf{b}$ must satisfy

$$0 = -\frac{\partial}{\partial \vec{p}} \cdot (\vec{a} f_0) + \frac{1}{2} \frac{\partial^2}{\partial \vec{p} \partial \vec{p}} : (\mathbf{b} f_0). \qquad (2.1.13)$$

If the background distribution function is not Maxwellian, then no general constraint on $\vec{a}$ and $\mathbf{b}$ is possible, because its deviation from thermodynamic equilibrium may maintain a stationary but nonequilibrium (and *a priori* unknown) form for f.

2.2 Charged Particle Equilibration

In this section I present explicit results for some processes of astrophysical interest by which nonequilibrium distribution functions relax to their equilibrium form, or by which individual test particles lose their initial momentum and energy, and come to be described by an equilibrium probabilistic distribution function. The complete calculations are generally rather lengthy, and numerous cases must be considered, so that I present abbreviated accounts of the derivations and summaries of the salient results. Rosenbluth, *et al.* (1957) present a particularly clear derivation of the theory of the Fokker-Planck coefficients for Coulomb (or gravitating) gases, Trubnikov (1965) contains a detailed discussion of this problem, and Spitzer (1962) provides a convenient summary of the explicit results required in many practical applications.

2.2.1 Dynamical Friction The simplest problem to consider is that of a nonrelativistic fast charged test particle slowing down as it moves through a background equilibrium plasma. The results calculated for this problem may be applied to gravitating neutral particles if the electrostatic potential $Z_1 Z_2 e^2/r$ is replaced by the gravitational potential $Gm_1 m_2/r$. This process is called Coulomb drag in a plasma, and dynamical friction in gravitating systems.

Consider a test particle with mass m_1, charge $Z_1 e$, and velocity $\vec{v} = v\hat{x}$ moving through a plasma consisting of randomly distributed "field" particles of mass m_2, charge $Z_2 e$, and mean number density n_2. All plasmas must contain more than one species of particle in order to maintain charge-neutrality, but the effects of the several species are almost exactly additive, so it is only necessary to calculate the effects of one species. The calculations are much simplified if the thermal velocities $\vec{v}'$ of the plasma particles may be neglected. This is a good approximation if their thermal energies are much less than the test particle kinetic energy.

The differential cross-section for classical Coulomb scattering (Eisberg 1961) is

$$\frac{d\sigma}{d\Omega} = \frac{Z_1^2 Z_2^2 e^4}{4m_{12}^2 u^4} \left[\sin(\theta/2)\right]^{-4}, \tag{2.2.1}$$

where $m_{12} \equiv m_1 m_2/(m_1 + m_2)$ is the reduced mass, θ is the scattering angle in the center of mass frame, and $u = |\vec{u}|$, where $\vec{u} \equiv \vec{v} - \vec{v}'$ is the relative velocity of the test particle with respect to the field particle from which it scatters. If $\vec{u}$ suffers a change $\Delta\vec{u}$ in a collision then the change in $\vec{v}$ is

$$\Delta\vec{v} = \frac{m_2}{m_1 + m_2}\Delta\vec{u}; \tag{2.2.2}$$

the remaining fraction $m_1/(m_1+m_2)$ of $\Delta\vec{u}$ appears as recoil velocity of the field particle.

Because the scattering is elastic u does not change as the result of a collision, but $\vec{u}$ is rotated by the scattering angle θ. If the scattering plane makes an angle φ to the x-y plane, then the components of $\Delta\vec{u}$ are readily found from trigonometry to be

$$\begin{aligned}
\Delta u_x &= -u(1-\cos\theta) = -2u\sin^2(\theta/2) \\
\Delta u_y &= u\sin\theta\cos\varphi = 2u\sin(\theta/2)\cos(\theta/2)\cos\varphi \\
\Delta u_z &= u\sin\theta\sin\varphi = 2u\sin(\theta/2)\cos(\theta/2)\sin\varphi
\end{aligned} \tag{2.2.3}$$

To compute the mean rate of change of velocity of the test particle, integrate over all scattering angles

$$\frac{d\langle\vec{v}\rangle}{dt} = \frac{m_2}{m_1+m_2}\int d\Omega\,\frac{d\sigma}{d\Omega}\,n_2 v\,\Delta\vec{u}(\Omega). \tag{2.2.4}$$

The factor $n_2 v$ is the rate at which the test particle encounters field particles, and is required by the normalization of $d\sigma/d\Omega$ to a single scatterer. It is evident that Δv_y and Δv_z average to zero when the integral over φ is performed, because the impulse perpendicular to the path of the incident test particle is equally likely to be in the $+\hat{y}$ direction as $-\hat{y}$, and in the $+\hat{z}$ direction as $-\hat{z}$. Then

$$\frac{d\langle\vec{v}\rangle}{dt} = -\hat{x}\frac{\pi Z_1^2 Z_2^2 e^4 n_2}{m_1 m_{12} v^2}\int_0^\pi \frac{\sin\theta\,d\theta}{\sin^2(\theta/2)}. \tag{2.2.5}$$

The indefinite integral of the integrand in (2.2.5) is $4\ln\sin(\theta/2)$, so that the definite integral diverges logarithmically at its lower limit. It is necessary to introduce a cutoff on the range of integration. The potential surrounding a static test charge q in a thermal equilibrium plasma is not the bare potential q/r, but is found (by combining Poisson's equation and the thermal equilibrium distribution of charges in a potential) to be

$$\phi(r) = \frac{q}{r}\exp(-r/\lambda_D), \tag{2.2.6}$$

where the Debye length

$$\lambda_D \equiv \sqrt{\frac{k_B T}{4\pi \sum_i n_i q_i^2}}. \tag{2.2.7}$$

The sum runs over all the species making up the plasma (this sum is the reason the effects of the various plasma species in slowing the test particle are not strictly additive). It is therefore sensible to cut off the divergent integration in (2.2.5) by making the lower bound of the integral the angle θ_{min}, at which the impact parameter of the collision equals λ_D. From an elementary impulse-approximation analysis this angle is found to be

$$\theta_{min} = \frac{2Z_1 Z_2 e^2}{m_{12} u^2 \lambda_D}. \tag{2.2.8}$$

Equation (2.2.5) then becomes

$$\frac{d\langle\vec{v}\rangle}{dt} = -\hat{x}\frac{4\pi Z_1^2 Z_2^2 e^4 n_2 \ln \Lambda}{m_1 m_{12} v^2}, \tag{2.2.9}$$

where Λ is the argument of the "Coulomb logarithm":

$$\Lambda = \frac{m_{12} v^2 \lambda_D}{Z_1 Z_2 e^2}. \tag{2.2.10}$$

The potential (2.2.6) is not strictly applicable to a moving charge, and a proper treatment requires consideration of the dynamic, rather than static, shielding properties of the plasma (Montgomery and Tidman 1964). The use of an abrupt cutoff is also not strictly correct; in principle $d\sigma/d\Omega$ could be calculated for the dynamically shielded potential. If either test or field particles are electrons then the cross-section should be calculated quantum mechanically (Spitzer 1962). Fortunately, a logarithm is very forgiving of such corrections, so long as its argument $\Lambda \gg 1$. In practice it is rarely necessary to be scrupulously careful in evaluating Λ. Values of $\ln \Lambda$ typically are in the range $20-30$ in interstellar plasmas, are ~ 10 in hot accretion flows and stellar winds, but may be $\lesssim 1$ in stellar interiors. Our derivation has implicitly assumed $\Lambda \gg 1$, and becomes inapplicable when this condition is not met. When it is applicable, neglected effects introduce fractional inaccuracies $\mathcal{O}(1/\ln \Lambda)$; these are referred to as non-dominant terms.

The logarithmic divergence of the integral in (2.2.5) means that each decade (or octave) of scattering angle θ between θ_{min} and π contributes equally to the slowing down of the test particle. Over nearly all of this range $\theta \ll 1$, so that most (all but a fraction $\sim 1/\ln \Lambda$) of the slowing down is a consequence of small angle collisions. Much of it is also a consequence of collisions with impact parameters $b \gtrsim n_2^{-1/3}$. Because in such a collision the acceleration of the test particle extends over a time $\sim b/u$ and distance $\sim b$, many such wide collisions take place simultaneously. As long as they cumulatively produce little change in $\vec{u}$, they may still be regarded as independent additive events, each described by the differential cross-section (2.2.1).

If we ignore the dependence of $\ln \Lambda$ on u (which is weak because it is logarithmic), then (2.2.9) is readily integrated to give the time t_s and length ℓ_s required for the test particle to stop:

$$t_s = \frac{m_1 m_{12} v^3}{12\pi Z_1^2 Z_2^2 e^4 n_2 \ln \Lambda} \tag{2.2.11}$$

$$\ell_s = \frac{m_1 m_{12} v^4}{16\pi Z_1^2 Z_2^2 e^4 n_2 \ln \Lambda}. \tag{2.2.12}$$

Numerical evaluation for a fast electron of energy E interacting with a hydrogen plasma, taking $\Lambda = 10$, yields

$$n_2 \ell_s \approx 2.0 \times 10^{17} \left(\frac{E}{1\ \mathrm{KeV}}\right)^2 \mathrm{cm}^{-2}. \tag{2.2.13}$$

For a fast proton

$$n_2\ell_s \approx 2.0 \times 10^{20} \left(\frac{E}{1 \text{ MeV}}\right)^2 \text{cm}^{-2}. \tag{2.2.14}$$

The rapid increase of stopping length with v and E is apparent. The stopping length is greater for an electron than for a proton of the same energy, but greater for a proton than for an electron of the same velocity, in each case by a factor $\sim m_p/m_e \sim 10^3$.

These results are inapplicable if the test particle is relativistic. Then ℓ_s is roughly proportional to energy for electrons. Energetic nucleons and nuclei lose their energy in violent nuclear collisions, whose cross-sections are roughly independent of energy, and which for protons exceed Coulomb slowing in importance for $E \gtrsim 100$ MeV.

Because the test particle is deflected as well as slowed in collisions, the stopping length ℓ_s is measured along its actual path. It may be seen by considering its deflections (**2.2.2**) that in most cases the path is approximately straight, at least until nearly all its energy has been lost, so that ℓ_s is a good approximation to the depth of penetration into the stopping material. The most important exceptions to this are fast electrons in a medium with $Z_2 \gg 1$, which randomize their directions before they lose much of their energy.

The slowing and energy loss of fast particles is predominantly the effect of field electrons. In a collision of given impact parameter a field ion (of $Z_2 = 1$) and electron will receive essentially the same impulse Δp, but the acquired kinetic energy $\Delta p^2/(2m_2)$ is much greater for the field electron. The passage of fast particles through cold matter predominantly heats the electrons. This effect is much reduced if the electron thermal velocity becomes comparable to the test particle speed (in which case it is no longer "fast"). The deflection of the test particles results, in comparable amounts (if $Z_2 = 1$), from the influence of field ions and field electrons; unless $Z_2 \gg 1$ deflection is usually unimportant.

A result very similar to (2.2.9) applies to the passage of fast particles through neutral matter. The lower cutoff on the integral in (2.2.5) must now be taken at collisions which impart enough energy to a bound electron to ionize it. For impacts significantly more violent than these the electronic binding is negligible and (2.2.1) is approximately correct. More distant encounters are ineffective, so that the appropriate value of $\ln \Lambda$ is less than it would be in an ionized medium.

The rare collisions with $\theta \sim 1$ are not completely insignificant; they contribute a fraction $\mathcal{O}(1/\ln \Lambda)$ of the total slowing. Because a fast particle undergoes only a few such collisions in the course of its slowing, their contribution is not accurately described by an integral over the differential cross-section, but varies from particle to particle. As a result the actual range of a particle may differ slightly from its mean value (2.2.12); this phenomenon is called straggling.

In cold dense matter evaluation of (2.2.10) for thermal particles $(\frac{1}{2}m_1 v^2 \approx \frac{3}{2}k_B T)$ implies $\ln \Lambda \lesssim 1$. This occurs in the interiors of low mass stars; at the Solar center $\ln \Lambda \approx 2$. When $\ln \Lambda$ is this small the theory is inapplicable. The plasma is "strongly coupled," meaning that the Coulomb energies of nearest neighbors are comparable to (or exceed) $k_B T$. The ions and electrons do not move freely, and the plasma is better described as a liquid than a gas. Stopping lengths of thermal particles are smaller than the mean interparticle separation $\sim n_2^{-1/3}$, and the theory of this section is inapplicable.

If this theory is to be applied to gravitating systems λ_D must be redefined. The phenomenon of Debye shielding does not exist because the particles cannot be in thermodynamic equilibrium (in equilibrium the most probably state would have particles with zero separation and infinite binding energy). If the substitution $q^2 \rightarrow Gm^2$ is made in (2.2.7), and the thermal velocity of the particles $\sqrt{k_B T/m}$ is taken to be the sound speed, then λ_D becomes the Jeans length λ_J (see **3.2**). We know that no stable self-gravitating configuration can exceed λ_J in size. A gas of point masses cannot be confined by external pressure, and will necessarily have a size close to λ_J. It is therefore customary and plausible to take the lower cutoff on the angular integration (2.2.5) in a gravitating system to be at angles for which the impact parameter is the size of the system; there are few field particles at greater distances.

A particularly simple result is obtained for gravitating systems in the limit $m_2 \ll m_1$. Substitution into (2.2.9) gives

$$\frac{d\langle\vec{v}\rangle}{dt} = -\hat{x}\frac{4\pi G^2 m_1 \rho \ln\Lambda}{v^2}, \tag{2.2.15}$$

where the mean background density $\rho = n_2 m_2$. This should be compared to the result (3.5.10) for the rate at which a mass moving through a fluid medium accretes additional mass, a process which also slows it. The slowing by dynamical friction, calculated in (2.2.15), exceeds the accretional slowing by a factor $\ln\Lambda$; this calculation of dynamical friction is applicable to supersonic motion through collisional fluids as well as to free particles. In a collisional fluid accretion occurs only from orbits whose deflection exceeds $2\sin^{-1}(1/\sqrt{5}) \approx 53°$, which represent only $\sim 1/\ln\Lambda$ of the integral in (2.2.5). The rate of slowing by dynamical friction is proportional to the test particle mass (the drag force and the accretion rate are proportional to its square), so that dynamical friction may be important for massive bodies, such as large molecular clouds, star clusters, and galaxies, even when it is insignificant for individual stars.

2.2.2 Equipartition Times There is another limit, complementary to that of a fast test particle moving through a cold plasma, in which it is easy to analyse the effects of Coulomb collisions on the velocity distribution function of test particles. Consider test particles of mass m_1 and charge $Z_1 e$ which are initially at rest ($\vec{v} = 0$) in a plasma of particles of mass m_2, charge Z_2e, and number density n_2, in thermal equilibrium at temperature T_2. As in **2.2.1**, the effects of several species are simply additive (except that they all contribute to the Debye length), so we consider only one.

Because initially $\vec{v} = 0$ and an equilibrium plasma is isotropic, there is no preferred direction, and we must have

$$\frac{d\langle\vec{v}\rangle}{dt} = 0. \tag{2.2.16}$$

However, Coulomb collisions do impart velocity to the test particle, so we can calculate the increase in its velocity dispersion $\langle v^2\rangle$. As long as the test particle may be considered to be nearly at rest the rate of increase of $\langle v^2\rangle$ is independent of the direction of the colliding field particle (again, because there is no preferred direction). For algebraic simplicity take the field particles to be travelling in the $-\hat{x}$ direction, so the results (2.2.3), together with (2.2.2), may be used to calculate the change of $\langle v^2\rangle$. Note that (2.2.3) would give erroneous results for each of $\langle v_x^2\rangle$, $\langle v_y^2\rangle$, and $\langle v_z^2\rangle$, but it gives the correct average (and sum $\langle v^2\rangle$), because taking this average is equivalent to averaging over all possible directions of impact of the field particles.

The rate of change of the velocity dispersion is then given by

$$\frac{d\langle v^2\rangle}{dt} = \left(\frac{m_2}{m_1+m_2}\right)^2 \int d\Omega\, d^3v_2\, \frac{d\sigma}{d\Omega} f(v_2)[\Delta u^2(\Omega, v_2)], \tag{2.2.17}$$

where the equilibrium distribution function is

$$f(v_2) = n_2 \left(\frac{m_2}{2\pi k_B T_2}\right)^{3/2} \exp\left(-\frac{m_2 v_2^2}{2k_B T_2}\right), \tag{2.2.18}$$

and from (2.2.3) we find $\Delta u^2 = 4u^2 \sin^2(\theta/2)$. Substituting (2.2.1), Δu^2, and $f(v_2)$ into (2.2.17) gives

$$\begin{aligned}\frac{d\langle v^2\rangle}{dt} = {} & \left(\frac{m_2}{m_1+m_2}\right)^2 \left(\frac{m_2}{2\pi k_B T_2}\right)^{3/2} \frac{8\pi^2 n_2 Z_1^2 Z_2^2 e^4}{m_{12}^2} \\ & \times \int \frac{\sin\theta\, d\theta}{\sin^2(\theta/2)} \int dv_2\, v_2 \exp\left(-\frac{m_2 v_2^2}{2k_B T_2}\right).\end{aligned} \tag{2.2.19}$$

The integral over θ is exactly the same integral evaluated in (2.2.5), and gives $4 \ln \Lambda$. Now Λ is defined for an encounter with the mean thermal energy $\frac{3}{2}k_B T$ (where T is an average of the test particle temperature, here taken to be zero, and T_2, weighted according to the reciprocals of m_1 and m_2 in order to give the mean kinetic energy of relative motion):

$$\Lambda = \frac{3k_B T \lambda_D}{Z_1 Z_2 e^2}. \tag{2.2.20}$$

Properly Λ varies with v_2, but this gives an insignificant correction (smaller than other neglected effects mentioned in **2.2.1**) to the final result. We now have

$$\frac{d\langle v^2\rangle}{dt} = \left(\frac{m_2}{m_1+m_2}\right)^2 \left(\frac{m_2}{2\pi k_B T_2}\right)^{3/2} \frac{32\pi^2 n_2 Z_1^2 Z_2^2 e^4 \ln \Lambda}{m_{12}^2} \times \int_0^\infty dv_2 \; v_2 \exp\left(-\frac{m_2 v_2^2}{2k_B T_2}\right). \tag{2.2.21}$$

This integral over velocity is elementary. Performing it, and collecting factors gives

$$\frac{d\langle v^2\rangle}{dt} = \frac{16\pi n_2 Z_1^2 Z_2^2 e^4 \ln \Lambda}{m_1^2} \sqrt{\frac{m_2}{2\pi k_B T_2}}. \tag{2.2.22}$$

We now define an equipartition time t_{eq} as the characteristic time for the mean kinetic energy to relax to the value $\frac{3}{2}k_B T$ it has if the test particles are in equilibrium with the field particles:

$$\begin{aligned} t_{eq} &\equiv \frac{\frac{3}{2}k_B T_2}{\frac{1}{2}m_1 \frac{d}{dt}\langle v^2\rangle} \\ &= \frac{3m_1 (k_B T_2)^{3/2}}{8\sqrt{2\pi m_2} n_2 Z_1^2 Z_2^2 e^4 \ln \Lambda}. \end{aligned} \tag{2.2.23}$$

The $3/2$ power of T appearing in this equation resembles the v^3 dependence in (2.2.11), and has a similar origin in the velocity dependence of the cross-sections.

These results are applicable only in the limit in which the test particles move slowly compared to the field particles; unless $m_1 \gg m_2$ this approximation will break down before equipartition is achieved. In the more general case in which the test particles are neither very fast (**2.2.1**) nor very slow (**2.2.2**) compared to the field particles, the kinematics of the collisions becomes more complex. Spitzer (1962) gives the result

$$t_{eq} = \frac{3m_1 m_2}{8\sqrt{2\pi} n_2 Z_1^2 Z_2^2 e^4 \ln \Lambda} \left(\frac{k_B T_1}{m_1} + \frac{k_B T_2}{m_2} \right)^{3/2} \tag{2.2.24}$$

for the equipartition time between a distribution of test particles which have an equilibrium distribution with temperature T_1 and field particles at temperature T_2. Equation (2.2.23) describes the case $T_1 = 0$. The result (2.2.24) may also be applied to the relaxation towards equilibrium of a distribution of particles with itself.

The parentheses in (2.2.24) correspond to the v^3 or $T^{3/2}$ dependence of (2.2.11) or (2.2.23), but indicate that the larger of the two thermal velocities is to be used. A very fast ion passing through a cool plasma principally gives up its energy to the plasma electrons, until it slows below the electron thermal velocity. If $v \lesssim (k_B T/m_e)^{1/2}(m_e/m_p)^{1/3} = (k_B T/m_p)^{1/2}(m_p/m_e)^{1/6}$ (which may still correspond to kinetic energies far exceeding $k_B T_2$), the test particle predominantly heats the plasma ions, because of the effect of a small $m_2 = m_e$ in the denominator of (2.2.24) when applied to its interaction with the plasma electrons.

2.3 Comptonization

Comptonization is the name given to the process by which electron scattering brings a photon gas to equilibrium. Because photons have a negligible cross-section for scattering by other photons (although for photons with frequencies ν_1 and ν_2 satisfying $h^2\nu_1\nu_2 \geq m_e^2c^4$ photon-photon pair production and subsequent annihilation have the effect of photon-photon scattering), they can only come to equilibrium by interaction with matter. In the laboratory the walls of an enclosure are available to absorb and re-emit the radiation, but in most astrophysical problems there are no walls and absorptive processes in the matter are often very slow. In hot fully ionized matter, the most frequent and important process is Compton scattering. In the limit that the electrons are infinitely massive (compared to $h\nu/c^2$) and slow-moving, Compton scattering would make the photon angular distribution isotropic, but would not affect the frequency spectrum. Because electrons are of finite mass and do have random thermal velocities, the photon spectrum may change as a result of scattering, and net energy transfer between photons and electrons may occur. The finite electron mass leads to an electron recoil as a result of Compton scattering, and tends, on average, to transfer energy from photons to electrons. The random electron velocities produce random Doppler shifts of the scatterers in the laboratory frame, which, on average, tend to increase the photon energy at the expense of the electrons. The former effect increases with photon frequency, and the latter with electron energy; if the photons have a thermal equilibrium distribution at the same temperature as that of the electrons, the two effects will balance.

The term Comptonization is used if the electrons are in thermal equilibrium at temperature T, and if both k_BT and $h\nu$ are much less than m_ec^2, where ν is the frequency of a typical photon. In this non-relativistic limit a number of powerful approximations are possible, and a differential equation of Fokker-Planck form for the time evolution of the photon occupation number $n(\nu)$, assumed isotropic, is obtained. This equation was first published by Kompaneets (1957), and is generally referred to as the Kompaneets Equation. The fully relativistic case is computationally much more complex, and cannot be reduced to a Fokker-Planck equation because the photon frequency shift $\Delta\nu$ upon scattering does not satisfy $\Delta\nu \ll \nu$.

Comptonization is likely to be important when the temperature is high and the density is low, so that matter is fully ionized, and absorption processes (generally proportional to the square of the particle density) are less important than scattering (proportional to the density). Comptonization occurs whenever photons and electrons in the same volume are described by different temperatures. It is of particular interest when the temperature of a low density electron gas much exceeds the temperature of a Planck function with the same energy density, and the absorption optical depth is low enough that the photon spectrum falls far below a Planck function at the electron temperature; typically some low energy ($h\nu \lesssim k_B T$) photons are present. Energy flows from the electrons to the photons. Three likely astrophysical sites have been suggested: the early universe, accretion flows into compact objects (such as in compact X-ray sources), and quasars. Comptonization has been used in theoretical modelling of each of these (though relativistic Compton scattering is probably more important for quasars). Comptonization will also occur when cool electrons are found in a radiation field of high energy ($h\nu \gg k_B T$), in which case the net energy transfer is from the photons to the matter.

Despite the wide use of Comptonization by theorists, no clear observational evidence for it exists. There can be no doubt that the process occurs, for it is derived from elementary relativistic kinetics, but its actual importance in astronomical objects is unproven. It produces no unique spectral signature, so it may be important without that fact being apparent. In some circumstances it may bound the parameters of observable objects; if Compton scattering and energy transfer are effective, the object may not be practicably observable because they may drain energy from an observable form, such as electrons which may emit synchrotron radiation, to a less observable form.

Kompaneets' original paper and the book by Rybicki and Lightman (1979) are rather terse, so I will try to explain the derivation in more detail. The elementary scattering process begins with an electron of momentum $\vec{p}$ and energy E, and a photon of frequency ν directed along the unit vector $\hat{n}$, and scatters these into $\vec{p}'$, E', ν', and $\hat{n}'$. Define the frequency shift $\Delta \equiv \nu' - \nu$. Conservation of energy and momentum state

$$h\nu + \frac{\vec{p}^2}{2m_e} = h\nu' + \frac{\vec{p}'^2}{2m_e} \tag{2.3.1}$$

$$\frac{h\nu}{c}\hat{n} + \vec{p} = \frac{h\nu'}{c}\hat{n}' + \vec{p}'. \tag{2.3.2}$$

Eliminate $\vec{p}'$ by regrouping and squaring (2.3.2) and substituting it into (2.3.1). Collecting terms and ignoring the term in Δ^2 (valid in the nonrelativistic limit) leads to a linear equation for Δ with the solution

$$h\Delta = -\frac{h\nu c\vec{p}\cdot(\hat{n} - \hat{n}') + h^2\nu^2(1 - \hat{n}\cdot\hat{n}')}{m_e c^2 + h\nu(1 - \hat{n}\cdot\hat{n}') - c\vec{p}\cdot\hat{n}'}. \tag{2.3.3}$$

Kompaneets (1957) contains typographical errors in the corresponding expression and elsewhere.

Now take $h\nu \sim k_B T \sim O(m_e v^2)$, where v is a characteristic electron thermal velocity. The second term in the numerator is an $O(v/c)$ correction to its first term, as is the third term in the denominator to its first term. The second term in the denominator is an $O(v^2/c^2)$ correction to its first term. We also see that $h\Delta/m_e c^2 \sim O(v^3/c^3)$. Had we included the Δ^2 term in deriving (2.3.3), it would have led to an $O(v^3/c^3)$ fractional correction, by adding an additional $h\Delta/2$ to the denominator; equivalently, the ratio Δ/ν which is $O(v/c)$ would have its value changed by an amount $O(v^4/c^4)$. This is clearly negligible. As will be explained later, we only need consider the leading term in (2.3.3), which is

$$h\Delta \approx -\frac{h\nu\vec{p}\cdot(\hat{n}-\hat{n}')}{m_e c}. \tag{2.3.4}$$

Now we must write down the Boltzmann (kinetic) equation governing the evolution of the photon occupation number $n(\nu)$. For an infinite homogeneous volume $(\partial/\partial\vec{x}) = 0$ and $\vec{F} = 0$, so that

$$\frac{\partial n(\nu)}{\partial t} = \int d^3p\; cd\sigma\left[n(\nu)\big(1+n(\nu')\big)N(E) - n(\nu')\big(1+n(\nu)\big)N(E')\right]. \tag{2.3.5}$$

$N(E)$ is the electron distribution function per unit momentum space and unit real space. Because the electrons are in thermal equilibrium $N(E)$ depends only on energy and not on direction. $d\sigma = (d\sigma/d\Omega)d\Omega$ is the infinitesimal element of cross-section for scattering a photon into the element of solid angle $d\Omega$; in this scattering $(\nu, \vec{p}) \rightarrow (\nu', \vec{p}')$. $d\sigma/d\Omega$ is the differential scattering cross-section. The first term in the brackets represents the scattering of photons from frequency ν to frequency ν' by electrons of energy E. Because photons are bosons the factor $\big(1+n(\nu')\big)$ represents the effect of stimulated scattering. The electrons are assumed nondegenerate $(N \ll 1)$, so the corresponding fermion factor $\big(1-N(E')\big)$ is ignored. The second term similarly represents scattering from ν' to ν. The argument of N in this term is $E' = E + h\nu - h\nu' = E - h\Delta$ because in this scattering the electron energy changes from E' to E. This term represents the process which is the exact inverse of that of the first term; it proceeds at a different rate only because the occupation numbers $n(\nu)$ and $N(E)$ are different from $n(\nu')$ and $N(E')$. The differential cross-section $d\sigma$ is the same, because if it were not an equilibrium gas of photons and electrons could drive itself away from equilibrium, violating the second law of thermodynamics. For a given ν and $\hat{n}$ the electron momentum $\vec{p}$ and the scattering angle Ω together determine ν' and $\hat{n}'$, so there is no additional integration over these variables. Such an integration could have been included, but then the kinematic constraints (2.3.1) and (2.3.2) would have introduced a Dirac-δ function, which would have made the integrals trivial.

It is easy to see that if a general Bose distribution

$$n(\nu) = \frac{1}{\exp(a + h\nu/k_B T) - 1} \tag{2.3.6}$$

is substituted into (2.3.5) and N is taken to be Maxwellian at the same temperature T then the right hand side of (2.3.5) vanishes, as it must by the second law of thermodynamics. It is necessary to introduce an arbitrary chemical potential $-ak_BT$ into $n(\nu)$ because Comptonization conserves photons; equilibrium is achieved only subject to this constraint. The photon distribution cannot relax to the Planck function ($a = 0$) by Compton scattering alone. The assumption that the electrons are fully relaxed and nondegenerate is generally well justified when Comptonization is of interest.

Equation (2.3.5) is an integrodifferential equation for $n(\nu)$, which is hard to solve, except by "brute force" numerical techniques. Such a solution gives little physical insight. The reason for this difficulty is that as it is written (2.3.5) relates $\partial n(\nu)/\partial t$ to the values of n at all other frequencies ν'; it is nonlocal in frequency. In the nonrelativistic limit the only ν' which contribute are close to ν, so the right hand side of (2.3.5) may be expanded in powers of $\Delta/\nu \ll 1$. The integrals may then be performed explicitly, leaving a differential equation which is easy to solve, and which is readily interpreted. It must be remembered that the Kompaneets equation is based on an expansion in powers of Δ/ν, which is $\mathcal{O}(v/c)$. This is not a very small ratio under conditions in which Comptonization is of interest; for $T = 10^8$ °K it is about 0.2. Although the Kompaneets equation is strictly correct in the nonrelativistic limit, it is not very accurate under the conditions in which it is usually used.

It will be necessary to expand (2.3.5) to terms of the second power in Δ. Use the Taylor expansions

$$n(\nu') = n(\nu) + \frac{h\Delta}{k_B T}\frac{\partial n}{\partial x} + \frac{1}{2}\left(\frac{h\Delta}{k_B T}\right)^2 \frac{\partial^2 n}{\partial x^2} + \cdots \quad (2.3.7)$$

$$N(E - h\Delta) = N(E)\left[1 + \frac{h\Delta}{k_B T} + \frac{1}{2}\left(\frac{h\Delta}{k_B T}\right)^2 + \cdots\right], \quad (2.3.8)$$

where a nonrelativistic Maxwellian $N(E)$ has been assumed and $x \equiv h\nu/k_B T$ is a convenient scaled frequency variable. Substitute these expressions into (2.3.5) and collect terms to obtain

$$\frac{\partial n(\nu)}{\partial t} = \frac{h}{k_B T}\left(\frac{\partial n}{\partial x} + n + n^2\right) I_1 + \frac{1}{2}\left(\frac{h}{k_B T}\right)^2 \left(\frac{\partial^2 n}{\partial x^2} + 2(1+n)\frac{\partial n}{\partial x} + n + n^2\right) I_2 + \cdots, \quad (2.3.9a)$$

where

$$I_1 \equiv \int d^3p \; d\sigma \; cN(E)\Delta \quad (2.3.9b)$$

$$I_2 \equiv \int d^3p \; d\sigma \; cN(E)\Delta^2. \quad (2.3.9c)$$

This has been written as a power series in Δ. The term in Δ^0 is zero, as it must be, for if $\Delta = 0$ no photon can change its frequency. The crucial step in reducing the integrodifferential equation has been the replacement of $n(\nu')$ and $N(E')$ by functions of $n(\nu)$, $N(E)$, and their derivatives, which have been brought out of the integrals. The only function of ν' left in the integrals is Δ, which has been simply expressed in terms of other variables by (2.3.4).

Examination of (2.3.4) shows why it was necessary to include the Δ^2 term in (2.3.9). For any $(\hat{n} - \hat{n}')$, and to lowest order in v/c, I_1 is proportional to $\int d^3p \; \vec{p} \cdot (\hat{n} - \hat{n}')N(E)$, which is zero because $N(E)$ is independent of direction (it is sufficient that the mean electron momentum be zero). Thus, to lowest order in v/c, I_1 is zero.

This must be so because this term would correspond to a systematic Doppler shift of $\mathcal{O}(v/c)$ in each scattering, which cannot be present for the assumed isotropic distribution of electrons. The nonzero contribution from I_1 depends on a better estimate of Δ than (2.3.4), and is smaller by a factor $\mathcal{O}(v/c)$. Equation (2.3.3) could be used to estimate Δ to higher order in v/c, but the integration of the resulting expression would be difficult. Fortunately (but not miraculously) the integration of the Δ^2 term in (2.3.9) is tractable, and makes the explicit integration of the Δ term unnecessary. In the Δ^2 term the first approximation (2.3.4) to Δ is sufficient.

Substitute (2.3.4) into the integral I_2 (2.3.9c) to obtain

$$\begin{aligned} I_2 &= \int d^3p\, c d\sigma\, N(E)\Delta^2 \\ &= \left(\frac{\nu}{m_e c}\right)^2 \int c d\sigma\, d^3p\, N(E)\left(\vec{p}\cdot(\hat{n}-\hat{n}')\right)^2. \end{aligned} \tag{2.3.10}$$

In the last integral write $\vec{p}\cdot(\hat{n}-\hat{n}') = p\,|\hat{n}-\hat{n}'|\cos\psi$, where ψ is the included angle between these two vectors. The quantity $|\hat{n}-\hat{n}'|^2$ depends on the scattering angle but not on $\vec{p}$, and may be removed from the integral over electron momentum space. Because the remaining factors in the integral are isotropic ($N(E)$ by assumption), the angular part of the momentum space integral is just the angular integral of $\cos^2\psi$, which is $4\pi/3$ for any polar axis (the direction of $\hat{n}-\hat{n}'$). Performing the integral over the direction of the electron momentum $\vec{p}$ gives

$$I_2 = \frac{1}{3}\left(\frac{\nu}{m_e c}\right)^2 \int c d\sigma |\hat{n}-\hat{n}'|^2 \int 4\pi p^2\, dp\, N(E) p^2. \tag{2.3.11}$$

The last integral in (2.3.11) is just n_e times $\langle p^2\rangle$, or $2m_e$ times the electron kinetic energy density. Because $N(E)$ is Maxwellian this is $3k_B T m_e n_e$. Then

$$I_2 = \left(\frac{\nu}{m_e c}\right)^2 k_B T m_e n_e c \int d\Omega \frac{d\sigma}{d\Omega}|\hat{n}-\hat{n}'|^2. \tag{2.3.12}$$

The exact differential cross-section for Compton scattering is given by the Klein-Nishina formula. Because we are taking the nonrelativistic limit we approximate it by the Thomson differential cross-section $d\sigma/d\Omega = \frac{1}{2}r_e^2(1 + (\hat{n} \cdot \hat{n}')^2)$, where $r_e \equiv e^2/(m_e c^2)$ is the classical electron radius. This introduces a fractional error $O(h\nu/m_e c^2) \sim O(v^2/c^2)$, which may be ignored because in general we are only considering the leading terms in expansions whose successive terms are in the ratio $O(v/c)$. Similarly, we ignore the Lorentz transformations between the laboratory and scattering electron's frames in evaluating $d\sigma/d\Omega$ and $d\Omega$, because in the nonrelativistic limit these frames become identical. A fully relativistic treatment would include these transformations, and is cumbersome. (2.3.12) becomes

$$I_2 = \left(\frac{\nu}{m_e c}\right)^2 k_B T m_e n_e c \int d\Omega \frac{1}{2} r_e^2 (1 + (\hat{n} \cdot \hat{n}')^2)|\hat{n} - \hat{n}'|^2. \quad (2.3.13)$$

Substitute $|\hat{n} - \hat{n}'|^2 = \hat{n}^2 + \hat{n}'^2 - 2(\hat{n} \cdot \hat{n}') = 2 - 2(\hat{n} \cdot \hat{n}')$, and integrate over the photon scattering angle $d\Omega$. Integrals over a sphere of odd powers of $\hat{n} \cdot \hat{n}' = \cos\theta$ are zero, while the integral of $\cos^2\theta$ is $4\pi/3$. Finally, rewrite the result in terms of the angle-integrated Thomson cross-section $\sigma_{es} = \frac{8}{3}\pi r_e^2$ to obtain

$$I_2 = 2\left(\frac{\nu}{m_e c}\right)^2 k_B T m_e n_e \sigma_{es} c. \quad (2.3.14)$$

Substituting (2.3.14) into (2.3.9) shows that the integral of Δ^2 contributes to $\partial n/\partial t$ a term proportional to $x^2(\partial^2 n/\partial x^2)$, while the integral of Δ does not. This result will permit the derivation of the Kompaneets equation without any further computation of integrals. The procedure is analogous to using (2.1.13) to determine the Fokker-Planck coefficient $\vec{a}$ from $\mathbf{b}$.

Compton scattering conserves photons. Therefore the occupation number $n(\nu)$ must satisfy a conservation law in three dimensional photon momentum space. For an isotropic photon distribution n depends only on the magnitude of the photon momentum, and such a law takes the form

$$\frac{\partial n}{\partial t} = -\frac{1}{x^2}\frac{\partial(x^2 j)}{\partial x}, \tag{2.3.15}$$

where j is the "current density" of photons. Equation (2.3.9a) contains a term equal to $\partial^2 n/\partial x^2$ times a function of x (but not of n). Because (2.3.15) describes the same function it must have the same form, so that any term in j proportional to $\partial n/\partial x$ cannot contain any other dependence on n. Hence j must be of the form

$$j(n,x) = g(x)\left(\frac{\partial n}{\partial x} + h(n,x)\right), \tag{2.3.16}$$

where g and h are functions yet to be determined.

The Bose equilibrium distribution function (2.3.6) must give zero when substituted into (2.3.9) (as it is a stationary solution of 2.3.5). Because there are no photon sources or sinks, $j = 0$ for a stationary solution. For n of the form (2.3.6)

$$\frac{\partial n}{\partial x} = -n - n^2 \tag{2.3.17}$$

holds identically. Therefore, the condition that $j = 0$ in equilibrium is satisfied if for all n and x

$$h(n,x) = n + n^2. \tag{2.3.18}$$

To determine g compare the coefficient of $\partial^2 n/\partial x^2$ in (2.3.15) and (2.3.16) to that in (2.3.9). In (2.3.9) it is x^2 times constants (the x^2 comes from the ν^2 in 2.3.14), so that $g(x) \propto x^2$. The constant factors are found from (2.3.9) and (2.3.14), so that

$$g(x) = -\frac{k_B T}{m_e c^2} n_e \sigma_{es} c x^2. \tag{2.3.19}$$

Combining (2.3.15), (2.3.16), (2.3.18), and (2.3.19), and defining a dimensionless scaled variable y in place of t

$$y \equiv t \frac{k_B T}{m_e c^2} n_e \sigma_{es} c \tag{2.3.20}$$

gives the Kompaneets equation:

$$\frac{\partial n}{\partial y} = \frac{1}{x^2} \frac{\partial}{\partial x} \left[x^4 \left(\frac{\partial n}{\partial x} + n + n^2 \right) \right]. \tag{2.3.21}$$

It is possible to determine I_1 explicitly, now that (2.3.21) and (2.3.14) are known, by comparing the coefficients of $\partial n / \partial x$. The result is

$$I_1 = \frac{k_B T}{m_e c^2} \sigma_{es} n_e x (4 - x). \tag{2.3.22}$$

This result may also be obtained by comparing the coefficients of n and n^2; the answers are necessarily the same.

In most astrophysically interesting Comptonization problems the photon density is far below its equilibrium value, so $n \ll 1$ and the nonlinear term in (2.3.21) may be neglected. Then it is possible to derive a simple result for the variation of the total photon energy density with time, if nearly all the energy is at low frequencies $x \ll 1$. This energy density is given by

$$\mathcal{E}_r = \frac{2(k_B T)^4}{h^3 c^3} \int_0^\infty n(x) x \, 4\pi x^2 dx. \tag{2.3.23}$$

Then

$$\begin{aligned} \frac{h^3 c^3}{8\pi (k_B T)^4} \frac{d\mathcal{E}_r}{dy} &= \frac{\partial}{\partial y} \int_0^\infty n x^3 \, dx \\ &= \int_0^\infty dx \, x^3 \frac{\partial n}{\partial y} \\ &= \int_0^\infty dx \, x \frac{\partial}{\partial x} \left[x^4 \left(\frac{\partial n}{\partial x} + n \right) \right] \\ &= \int_{x=0}^\infty x \, d \left[x^4 \left(\frac{\partial n}{\partial x} + n \right) \right], \end{aligned} \tag{2.3.24}$$

where the n^2 term in (2.3.21) has been neglected. Integrate by parts, and assume that $n(x)$ drops off sufficiently rapidly (it will usually decline exponentially) as $x \rightarrow \infty$, but does not rise too rapidly as $x \rightarrow 0$, so that $x^5 \left(\frac{\partial n}{\partial x} + n\right) \rightarrow 0$ in both these limits. Then

$$\begin{aligned} \frac{h^3c^3}{8\pi(k_BT)^4}\frac{d\mathcal{E}_r}{dy} &= -\int_0^\infty x^4\left(\frac{\partial n}{\partial x} + n\right)\, dx \\ &= -\int_0^\infty nx^4\, dx - \int_0^\infty x^4\frac{\partial n}{\partial x}\, dx. \end{aligned} \tag{2.3.25}$$

Again integrating by parts,

$$\frac{h^3c^3}{8\pi(k_BT)^4}\frac{d\mathcal{E}_r}{dy} = -\int_0^\infty nx^4\, dx + 4\int_0^\infty nx^3\, dx. \tag{2.3.26}$$

If the photon energy density is concentrated at low frequencies $x \ll 1$, the first integral in (2.3.26) is much less than the second. Neglect the first integral to obtain the approximate result

$$\frac{d\mathcal{E}_r}{dy} = 4\mathcal{E}_r \tag{2.3.27}$$

or

$$\mathcal{E}_r = \mathcal{E}_0 \exp(4y). \tag{2.3.28}$$

The energy density grows exponentially (even though photon number is conserved). The e-folding time is

$$t_{C\gamma} = \frac{t}{4y} = \frac{m_ec^2}{4k_BT}\frac{1}{n_e\sigma_{es}c}. \tag{2.3.29}$$

The characteristic time scale (2.3.20 and 2.3.29) is proportional to $(n_e\sigma_{es}c)^{-1}$, a photon's mean time between Compton scatterings, and inversely proportional to k_BT/m_ec^2, because the mean photon energy gain per scattering is $O(k_BT/m_ec^2) \sim O(v^2/c^2)$.

The exponential growth indicated by (2.3.28) does not continue indefinitely. Eventually a significant fraction of the photon energy comes to be in photons for which $x \gtrsim 1$. Then the negative term in (2.3.26) is no longer insignificant, and the growth slows.

A realistic astronomical problem is unlikely to start at some initial time with a soft photon spectrum whose energy density then grows exponentially according to (2.3.28). It is more plausible to think of an object of finite size and finite optical depth with a steady source of low frequency photons. As the photons increase their mean energy according to (2.3.28) they also diffuse outward and escape. A steady state is obtained if there is a steady heat source for the matter to replenish the energy it gives up to the photons. It is straightforward to compute numerically the emergent spectrum. Qualitatively, it depends on the ratio of the size of the Compton scattering cloud to a critical size approximately equal to $\sqrt{m_e c^2/k_B T}(n_e \sigma_{es})^{-1}$. For significantly larger clouds the energy growth saturates, and the emergent spectrum resembles a Wien law $n(\nu) \propto \exp(-h\nu/k_B T)$ (the low n or large a limit of the equilibrium Bose distribution 2.3.6). Smaller clouds are inefficient energy multipliers, and produce nearly power law spectra, steeply decreasing with increasing frequency. Because the critical size depends on T, the emergent radiative power is a steeply increasing function of T, and wide variations in the photon source or the power supplied to the matter are accommodated by modest changes in the steady state temperature.

In some circumstances the effect of Comptonization on the electrons is of more interest than its effect on the photons. This is a much simpler problem; because we have assumed that the electron distribution function quickly relaxes to a Maxwellian of temperature T, we only need to know the mean rate of energy transfer, and not the effect of Compton scattering as a function of electron momentum. The complete answer in the limit $n(\nu) \ll 1$ is supplied by (2.3.26), noting that each erg supplied to the photons is drawn from the electron thermal energy.

The energy transfer may be calculated explicitly for special cases of $n(\nu)$. For a general Bose distribution the integrals must be expressed as infinite series, and if $a \lesssim 1$ the approximation $n(\nu) \ll 1$ is not valid throughout the spectrum. In astrophysics one often deals with diluted black body radiation fields, in which the spectral shape resembles that of a black body but the energy density is much lower. The dilution is generally the result of spherical divergence, and is found in regions illuminated by a small distant source. Dilution reduces the energy density, but does not change $n(\nu)$ for photon states whose momenta are directed from the source; other states have $n = 0$, giving an anisotropic n. It is easier and usually a good approximation to consider an isotropic Wien spectrum of arbitrary intensity characterized by a temperature αT; then $n(\nu) = n_0 \exp(-x/\alpha)$. Insertion of this into (2.3.26) (which implicitly assumes $n \ll 1$) leads to the result

$$\frac{d\mathcal{E}_e}{dt} = \frac{8}{3}\mathcal{E}_e \frac{\mathcal{E}_r \sigma_{es}}{m_e c}(\alpha - 1), \tag{2.3.30}$$

where $\mathcal{E}_e$ is the electron thermal energy density. For very soft photon spectra $(\alpha \ll 1)$ the electrons cool at the rate

$$\frac{d\mathcal{E}_e}{dt} = -\frac{8}{3}\mathcal{E}_e \frac{\mathcal{E}_r \sigma_{es}}{m_e c}. \tag{2.3.31}$$

(2.3.31) may be obtained regardless of the shape of the photon spectrum by considering the Thomson scattering drag force on the thermal motion of the electrons, allowing for Doppler shifts and aberration to first order in v/c; for $h\nu \ll k_B T$ the increase in electron velocity dispersion resulting from scattering recoil is negligible. These results hold for any angular distribution of radiation, provided the mean radiation pressure has been subtracted out and the electron distribution remains isotropic.

An example of the application of (2.3.30) and (2.3.31) is to the accretion of matter onto a white dwarf or a neutron star. The stellar photosphere is a source of radiation with temperature

$T_e \approx (L/4\pi r^2 \sigma_{SB})^{1/4}$, where L is the luminosity and σ_{SB} is the Stefan-Boltzmann constant; this may typically be $\sim 3 \times 10^5$ °K for a white dwarf and $\sim 10^7$ °K for a neutron star. If matter falling freely from infinity is stopped in a shock at the stellar surface, its temperature is of order $T \sim (GM\mu/k_B R)$, where μ is the molecular weight; this is $\sim 10^9$ °K for a white dwarf and could range as high as $\sim 10^{12}$ °K for a neutron star (though a simple accretion shock is not expected). Thus very hot matter is immersed in a radiation field of much lower color temperature, and (2.2.31) may be used to calculate the Compton cooling of the matter (provided the electron temperature satisfies $k_B T \ll m_e c^2$).

From (2.3.31) we see that if no heat sources are present the electron thermal energy decays exponentially on a characteristic Compton cooling time

$$t_{Ce} = \frac{3m_e c}{8\mathcal{E}_r \sigma_{es}}. \tag{2.3.32}$$

If $\mathcal{E}_r = L/(4\pi r^2 c)$, corresponding to a soft photon luminosity L flowing radially outward, then

$$t_{Ce} = \frac{L_E}{L}\frac{3m_e}{8m_H}\frac{r^2 c}{GM}, \tag{2.3.33}$$

where L has been expressed in terms of the Eddington luminosity L_E (1.11.6) for pure hydrogen composition and m_H is the mass of a hydrogen atom. This cooling time should be compared to the hydrodynamic time t_h (1.6.1), which characterizes free fall or periods of vibration. Their ratio is

$$\frac{t_{Ce}}{t_h} = \frac{L_E}{L}\frac{3m_e}{8m_H}\sqrt{\frac{rc^2}{GM}}. \tag{2.3.34}$$

It is evident that if L is large $t_{Ce} \ll t_h$, so that Compton cooling is very rapid. It may dominate the energy balance in accretion flows onto compact objects, and therefore may determine material temperatures and the spectrum of observable radiation.

2.4 Evolution and Collapse of Star Clusters

There are two kinds of astronomical objects which have led to the study of the evolution and collapse of clusters of stars. The first of these is the globular clusters. These spectacularly beautiful objects typically contain 10^5 stars in a region perhaps 10 parsecs across; the central density of stars may be as much as 10^6 times higher than that in the Solar neighborhood. Only a small fraction of these stars appear in a photograph of a globular cluster, for they span a wide range of brightness and most are too faint to detect. Frequently the inner parts of a cluster resemble a single over-exposed blob of overlapping stellar images. Despite this, even the densest inner regions are quite empty; an observer at the center would see only $\sim 10^{-11}$ of the sky covered with stars. The escape velocity and random stellar velocity of globular clusters are quite small and difficult to measure, even though their central densities are high, because the total mass is modest. Typical velocities are probably around 10 km/sec, which should be contrasted to velocities of 250 km/sec in galaxies. Globular clusters are dynamically fragile objects.

The stars in globular clusters in our Galaxy are among the oldest known. (The Magellanic Clouds contain globular clusters, apparently recently formed, in which the stars are young.) Astronomers have been interested in globular clusters for nearly a century because they have been very useful in understanding stellar evolution and the extragalactic distance scale, and because some have hoped that they might illuminate the history of the early universe and the formation of our Galaxy. Interest in globular clusters surged in 1975, when it was discovered that they are about 100 times richer in compact X-ray sources than our galaxy as a whole. It was then widely speculated that globular clusters might contain 1% of their mass in a massive central black hole. This hypothesis is now largely discounted because the X-ray emission appears to be more characteristic of accreting neutron stars in binary systems (**4.2**; most globular cluster X-ray sources show characteristic bursts; Lewin and Joss 1983), and because the X-ray sources are not found in the exact cluster centers, implying that they are not very massive (Grindlay, *et al.* 1984). The overabundance of X-ray sources, and even their presence at all, are still unexplained. This problem is part of the more general problem of binary stars in globular clusters, about which there are few data, but which may be important for the dynamical evolution of the clusters.

The second astronomical application of the theory of star clusters is to elliptical galaxies, and possibly to quasars (**4.7**). Elliptical galaxies typically contain 10^{11} stars in a region 10^4 parsecs across. The stars themselves resemble those in globular clusters but they are distributed at a much lower density, and with much higher velocities (typically 250 km/sec). The evolutionary and collapse phenomena which are believed to be important for globular clusters are negligibly slow for elliptical galaxies. Interest lies in the possibility that some galaxies (either elliptical or spiral) contain much denser cores, intermediate in parameters between globular clusters and the galaxies as a whole, in which these phenomena may be important. There is evidence that some nearby galaxies (for example, Andromeda) have such cores, for photographs underexposed in order to study their inner regions show very compact nuclei with nearly star-like images (see also Light, *et al.* 1974). I will call these "pseudo-stellar nuclei;" the word "quasistellar" generally refers to quasars, which are very different (though possibly related) objects.

The fundamental problem in the theory of star clusters (reviewed by Lightman and Shapiro 1978) is the inability of the individual stars, considered as particles constituting a gas, to come to full thermodynamic equilibrium. Because the depth of the potential well is finite, in any Maxwellian distribution of stellar velocities there will always be a small fraction of the stars moving faster than the escape velocity. It is possible to estimate this fraction if one uses the virial theorem to relate the random thermal velocity to the mean gravitational potential of the cluster. The quantitative result depends on how the structure is modeled (for example, how the mean radius is defined in estimating the depth of the potential); typically, about 1% of the stars in a Maxwellian would have enough energy to escape. In practice, the stellar distribution function is cut off at the escape energy. As the distribution function collisionally relaxes towards a complete Maxwellian, there is a steady efflux of stars with positive energy. The remainder of the cluster contracts as it loses mass and its total binding energy grows in magnitude. The loss of mass is more important, and would imply a deepening of the gravitational potential well even were no energy lost to evaporation.

This fundamental problem of cluster thermodynamics also appears if one considers the equilibrium distribution of the orbital parameters of a binary star in thermal contact with a heat bath (the cluster) at temperature T. In equilibrium, the probability of being in a state of energy E is proportional to $\exp(-E/k_BT)$. The energy $E = -Gm_1m_2/(2a)$, where the stars have masses m_1 and m_2, and a is the semi-major axis of the orbit of their separation vector. As $a \to 0$, $E \to -\infty$, and the probability diverges. It is not possible to sum over these probabilities to obtain a partition function. Thermodynamic equilibrium is not possible, for essentially all the equilibrium probability density is in states of infinite binding energy. In practice, a binary star would not rapidly contract to an orbit of infinitesimal size, because as its orbit becomes smaller, its rate of relaxation by encounters with the stars making up the thermal bath rapidly decreases. The distribution of orbital parameters is always determined by the rates of relaxation processes, and never by a thermal equilibrium distribution function. The atomic version of this equilibrium catastrophe (aggravated by the rapid radiation of an orbiting classical electron) was a fundamental problem of physics resolved by the quantization of atomic states. Discs of matter orbiting a central object (**3.6**) present an analogous problem; a dissipative process (viscosity) draws matter into the central gravitational potential well, as the distribution of matter relaxes towards an (unattainable) equilibrium.

The inner parts of a globular cluster may be compared to the hypothetical binary star, with the outer parts representing the heat bath. Essentially all the equilibrium probability is in states which are "down the black hole" of infinitely tight binding. The cluster never actually comes to equilibrium, and the calculation of its properties requires the calculation of its nonequilibrium relaxation processes.

The collisional relaxation time for gravitating masses, such as stars in a cluster, may be calculated in essentially the same way as for charged particles in a plasma (**2.2**). The stars themselves only rarely collide, but their paths are changed by their mutual gravitational interaction. In carrying out the integration over impact parameters the upper cutoff is not the Debye length, for there is no Debye shielding in gravitational interactions (in the unattainable state of complete thermodynamic equilibrium there would be anti-shielding). Instead, the integration is cut off at the geometrical size of the cluster, or that of the dense core region in which collisional relaxation is important. The result (Lightman and Shapiro 1978) for clusters all of whose stars have the same mass is a characteristic relaxation time

$$
\begin{aligned}
t_r &= \frac{v_m^3}{15.4 G^2 m^2 n \ln(0.4N)} \\
&\approx 7 \times 10^8 \text{ yr} \left(\frac{N}{10^5}\right)^{1/2} \left(\frac{m}{M_\odot}\right)^{-1/2} \left(\frac{R}{5 \text{ pc}}\right)^{3/2},
\end{aligned}
\tag{2.4.1}
$$

which has been evaluated at the radius R which includes half of the cluster mass; in the logarithm N was taken equal to 10^5. In (2.4.1) v_m is the dispersion of the stellar velocities, m the stellar mass, n the mean density of stars inside of R, and N the total number of stars in the cluster. The coefficient differs somewhat from that obtained in **2.2** because an attempt has been made to allow for the nonuniformity of the cluster. The argument of the logarithm may be obtained from (2.2.10), replacing $Z_1 Z_2 e^2$ by Gm^2, λ_D by R, and using the virial theorem to estimate the velocity dispersion; the (uncertain) coefficient of N depends on the quantitative structure of the cluster.

The most important single implication of (2.4.1) is its order of magnitude. For a globular cluster the factors in parentheses are of order unity, so that the relaxation time is much shorter than the age of the cluster (globular cluster ages are estimated from stellar evolutionary arguments to be about 10^{10} years, close to the age of the Galaxy). Dynamical evolution is important for globular clusters. In contrast, for elliptical galaxies $N \sim 10^{11}$, $R \sim 10{,}000$ pc, and t_r is more than 10^{16} years; dynamical evolution is completely insignificant, unless there is an inner region with much higher density. Such a region must have n much greater than that of a globular cluster because of the factor v_m^3 in (2.4.1), which is much larger in elliptical galaxies. There would be no reason to suspect the existence of such a region, were it not for the unexpected existence of quasars, other non-stellar activity in galactic nuclei, and a few observed pseudo-stellar nuclei.

There is a second time scale of interest for a globular cluster, the dynamical time t_d required for a star to cross the cluster:

$$t_d \equiv R/v_m. \tag{2.4.2}$$

There is another significance to t_d. On time scales much shorter than t_r collisions may be ignored, and the Boltzmann equation (2.1.9) for the stellar distribution function f becomes the Vlasov equation (2.1.4) (closely related to the Vlasov equation used in the theory of collisionless plasmas):

$$\frac{df}{dt} = \frac{\partial f}{\partial t} + \vec{v} \cdot \frac{\partial f}{\partial \vec{r}} - \nabla\Phi \cdot \frac{\partial f}{\partial \vec{v}} = 0, \tag{2.4.3}$$

where Φ is the cluster gravitational potential. Some distribution functions imply $\partial f/\partial t = 0$, but some do not. If not, then the characteristic time scale on which f changes is of order R/v or $v/\nabla\Phi$, each of which is of order t_d. This rapid change in f is called violent relaxation.

There is no complete criterion, other than (2.4.3) itself, for the occurrence of violent relaxation. Thermodynamic equilibrium guarantees its absence, but it will not occur for many nonequilibrium f. Because f is a function of 6 variables (aside from time) it may be very complex. Several well-known examples of nonequilibrium distribution functions which do not undergo violent relaxation exist. For example, begin with a spherically symmetric cluster in equilibrium, with the distribution function isotropic and Maxwellian everywhere (except for the missing positive-energy tail). Define an arbitrary axis through the cluster center. Then, for each particle with positive azimuthal velocity about that axis, reverse the sign of the azimuthal component of its velocity. This bizarre distribution function will not undergo violent relaxation. $\nabla\Phi$ has no azimuthal component, because the density is spherically symmetric. $\partial f/\partial\vec{v}$ is not changed by the velocity reversal, except for its azimuthal component, so $\nabla\Phi \cdot (\partial f/\partial\vec{v})$ is unaffected by the velocity reversals. Similarly, $\partial f/\partial\vec{r}$ has no azimuthal component, while it is only the azimuthal component of $\vec{v}$ which changes, so $\vec{v} \cdot (\partial f/\partial\vec{r})$ is likewise unaffected. Hence, if $\partial f/\partial t = 0$ before the reversal, it is still so afterwards. It is even more remarkable that the cluster remains spherically symmetric despite having a large angular momentum. Only on the longer collisional time scale t_r will f change, and the cluster shape will flatten. Recall that for elliptical galaxies t_r is extremely long. This has led to the suggestion that the shapes of elliptical galaxies, in contrast to those of stars, planets, and globular clusters, may not be simply related to their angular momenta. Another example is a disc-like (but non-rotating) distribution of stars with a large isotropic velocity dispersion in the disc plane, but negligible velocities perpendicular to its plane. There may also be bar-like structures with one-dimensional velocity dispersions, and superpositions of two or three orthogonal bars or discs.

When violent relaxation occurs its consequences are not simple. It is natural to assume that the distribution function acquires a fine-grained structure in phase space as a result of rapid "scrambling," and that if one averages over this structure the resulting coarse-grained distribution function will then satisfy the Vlasov equation with $\partial f/\partial t = 0$. This averaging is reminiscent of the turbulent mixing of two diffusionless fluids, or of Gibbs's explanation of the entropy of mixing. Because no thermodynamic principle determines the end point of violent relaxation, it is not possible to specify it in advance. Violent relaxation is therefore very different from collisional relaxation, which we believe rapidly leads to a Maxwellian distribution. The latter belief is founded on the knowledge that a Maxwellian is a stationary solution to the Boltzmann equation, and is the solution of highest entropy. Except in pathological cases (for example, when one degree of freedom is completely uncoupled) it is probably the only stationary solution. No such governing principle is applicable to violent relaxation. Fortunately, it is rapid, and therefore is feasible to calculate numerically; some results are reviewed by Lightman and Shapiro. One consequence of the absence of a thermodynamic principle is that the endpoint may preserve a memory of the initial conditions in complex and subtle ways. Observation of a system which has undergone violent, but not collisional, relaxation, such as an elliptical galaxy, may give interesting information about its formation.

The ratio of the dynamical to the collisional time scale may be calculated from (2.4.1) and (2.4.2):

$$\frac{t_d}{t_r} \approx \frac{26 \log_{10}(0.4N)}{N}. \tag{2.4.4}$$

The number of stars N is a measure of the smoothness of the gravitational potential, and of the validity of the separation of violent (collective) and collisional (microscopic) relaxation time scales. In order that this separation be valid it is necessary that $N \gg 100$. Clusters of fewer members are best regarded as complex multiple stars. This includes many open clusters and clusters of galaxies (where the fact that galaxies are not point masses adds a further complication). Clusters of many members have a clear separation of time scales, which permits the simplification of calculating each process separately. Once violent relaxation is over, it is possible to calculate the effects of collisional relaxation by assuming that the particles move in a static potential, in orbits of nearly constant energy (and nearly constant angular momentum, if the potential is azimuthally symmetric).

Because direct measurements of the velocities of individual stars in globular clusters are difficult, most of our understanding of the structure and evolution of clusters is based on theory. Although the theory may be, and has been, tested against counts of stars, measurements of stellar velocities, or the distribution of light in clusters, they do not test it completely. Like all theoretical "understanding" of physical phenomena, it is subject to later empirical revision. It may be more secure than our similar "understanding" of stellar structure and evolution, because cluster dynamics is founded only upon Newtonian mechanics. Take this last sentence skeptically; I might not have written it had I not known of the unexpected result of the solar neutrino experiment.

The innermost core region of a globular cluster is isothermal, because the stars within it undergo rapid relaxation of their distribution function and are spatially mixed. Because the density is highest here, relaxation is even more rapid than indicated by (2.4.1), which is an average. Poisson's equation for the gravitational potential Φ in such a spherically symmetric region is

$$\frac{1}{r^2}\frac{d}{dr}\left(r^2\frac{d\Phi}{dr}\right) = 4\pi G\rho. \qquad (2.4.5)$$

Using a thermal velocity v_{th} to describe the stellar velocity distribution, the equilibrium density $\rho(r)$ may be written

$$\rho(r) = \rho_0 \exp\left(-\Phi(r)/v_{th}^2\right). \qquad (2.4.6)$$

Then define the dimensionless potential φ and radius ξ

$$\varphi \equiv \Phi/v_{th}^2 \qquad (2.4.7)$$

$$\xi \equiv r\sqrt{4\pi\rho_0 G/v_{th}^2}, \qquad (2.4.8)$$

and use (2.4.6) to rewrite (2.4.5):

$$\frac{1}{\xi^2}\frac{d}{d\xi}\left(\xi^2\frac{d\varphi}{d\xi}\right) = \exp(-\varphi). \qquad (2.4.9)$$

Equation (2.4.9) describes the isothermal core region of all relaxed star clusters, and need only be numerically integrated once, just like the Lane-Emden equation for polytropes of a given polytropic index (**1.10**). Its derivation would take exactly the same form for an isothermal collisional gas in hydrostatic equilibrium, so that it is also the Lane-Emden equation for an isothermal star.

From Earth we cannot directly measure the density $\rho(r)$. We can measure the density $\varrho(r)$ projected along a line of sight which passes a distance r from the cluster center. By elementary geometry

$$\varrho(r) = 2\int_r^\infty \rho(r')\frac{r'\,dr'}{\sqrt{r'^2 - r^2}}. \qquad (2.4.10)$$

Given a tabulated $\rho(r)$ the calculation of $\varrho(r)$ is simple. Less easy is the inverse problem, of inverting an observed $\varrho(r)$ to obtain $\rho(r)$.

This requires the solution of (2.4.10) as an integral equation for the unknown $\rho(r)$. If one replaces the integral by a finite sum then $\rho(r)$ (defined only at a finite number of points, as $\varrho(r)$ is observed) may be determined from a set of linear equations, solved by matrix inversion. Unfortunately, inverting a matrix of data, containing observational error, is much more treacherous than inverting a matrix of exactly known numbers. Observations of the inner parts of globular clusters are at least consistent with the isothermal core model.

Outside the isothermal core is a region known as the halo. This is roughly defined by the condition that collisional relaxation is very slow, because the density is lower than in the core. Two kinds of stellar orbits should be distinguished. Those of high angular momentum never enter the core, and play no part in its dynamical evolution. Stars may have entered such orbits at an earlier epoch in the evolution of the cluster, or been born in them. Orbits of low angular momentum enter the core, and undergo dynamical relaxation there, though most of life of a star in such an orbit is spent in the halo. These orbits may be considered to be the highest energy part of the core's distribution function, having nearly enough energy to escape the cluster entirely. The mass contained in the halo is not large, so that the potential may be considered to vary as $1/r$.

Stars in halo orbits have energies negative in sign, but of small absolute magnitude. As the cluster core contracts, stars diffuse through this region of near-zero energy, with a net flux towards $E = 0$. At $E = 0$ there is a sink in energy space, as stars freely escape the cluster. The density of stars in nearly-zero energy orbits may be estimated roughly but simply (following Lightman and Shapiro). For an isotropic distribution function the Fokker-Planck equation (2.1.10), in which the variables are t, $\vec{x}$, and $\vec{p}$, may be transformed by a change of variable into a simpler Fokker-Planck equation in which the variables are t, r, and E. The diffusion coefficient in energy varies smoothly through $E = 0$, and may be regarded as nearly constant in a small interval around $E = 0$. There is nothing "special" about this energy when a star is in the cluster core, where nearly all its dynamical relaxation takes place; relaxation depends on the kinetic energy $E - m\Phi$. The "specialness" is only present in the behavior of the potential as $r \to \infty$. The diffusion term in the Fokker-Planck equation dominates the dynamical friction term because of the abrupt cutoff of the distribution function at $E = 0$; near such a cutoff second derivatives are much larger than first derivatives. Therefore, an energy-independent flux of stars toward zero energy and escape requires that in the cluster core $\partial f/\partial E$ be constant for small $|E|$, or

$$f(E) \propto -E. \tag{2.4.11}$$

A star with energy E has an orbit which extends to a radius $r \propto |E|^{-1}$, and which has a period proportional to $r^{3/2}$ or $|E|^{-3/2}$, for the orbit outside of the cluster core is nearly Keplerian. All low angular momentum halo stars move through the cluster core at nearly the same speed, so the fraction of their lifetime spent in the core is proportional to $|E|^{3/2}$. In order to calculate the total number of stars per unit energy $N(E)$ the cluster contains, the density in the core $f(E)$ must be divided by the fraction of its life a star spends in the core, so that (2.4.11) implies

$$N(E) \propto |E|^{-1/2}. \tag{2.4.12}$$

We want $N(r)$, the number of stars whose orbits extend to radius r, per unit radius. Most of the stars found at r have orbits whose major axes are comparable to r. Then

$$N(r)\,dr = N(E)\frac{dE}{dr}dr \propto |E|^{3/2}\,dr \propto r^{-3/2}\,dr. \tag{2.4.13}$$

These stars are spread over a volume $\sim 4\pi r^2\,dr$, so their volume density $n(r)$ is

$$n(r) \equiv \frac{N(r)\,dr}{4\pi r^2\,dr} \propto r^{-7/2}. \tag{2.4.14}$$

Observation agrees with this theoretical estimate. This confirms the assumption that most of the stars in the halo have low angular momentum orbits, and enter these orbits as a result of the diffusion of core stars to zero energy and escape. Higher angular momentum stars would not undergo significant dynamical evolution, and their density would depend on their orbital parameters when formed; they need not follow (2.4.14).

The halo is cut off at a finite radius by the tidal effects of the Galactic gravitational field. Beyond that radius, the only stars are those which are freely escaping, or background ("field") stars accidentally encountering the cluster.

As stars escape the cluster the total binding energy of those left behind must increase. Because the mass left behind is decreasing, it must contract into an ever-smaller volume, at ever-increasing density. A proper calculation of this process requires integrating a Fokker-Planck equation, but a much simpler calculation reveals its qualitative nature. Assume that the escaping stars carry away zero energy; this approximation is rather good, and leads only to a slight underestimate of the rate of cluster evolution. Use (2.4.1) to estimate the cluster relaxation time, ignoring the variation of the logarithm, and use the virial theorem to estimate the velocity dispersion. Conservation of total cluster energy implies

$$R \propto N^2, \tag{2.4.15}$$

while the virial theorem implies

$$v^2 \propto \frac{N}{R} \propto N^{-1}. \tag{2.4.16}$$

Then the variation of N is given by

$$\frac{dN}{dt} \sim \frac{N}{t_r} \propto N^{-5/2}. \tag{2.4.17}$$

This may be integrated to give

$$\frac{N}{N_0} = \left(1 - \frac{t}{t_0}\right)^{2/7}. \tag{2.4.18}$$

The cluster evaporates entirely in a finite time t_0, except for an infinitesimal fraction of its mass which contains the full initial binding energy. More detailed calculations show that t_0 is typically 10 to 30 times the initial relaxation time t_r. In the course of this evaporation the density and velocity dispersion diverge as the radius contracts to zero.

This result is startling and intriguing, particularly because estimates of core relaxation times in observed globular clusters are as low, in some cases, as 10^7 years. These clusters should collapse soon, and, unless we are at a "special" and preferred cosmological moment, others have collapsed in the past. What actually happens, and what does the remnant look like afterward?

One obvious oversimplification we have made has been to treat the integer N as a continuous variable. If a cluster is reduced to a single very tightly bound binary star ($N = 2$), evolution stops. To absorb all the binding energy of a cluster of N stars, such a binary would have to have an orbit whose size is $\sim N^{-2}$ times the initial core radius. This is not possible for ordinary stars, which are too large, but a binary made of degenerate stars or black holes could be sufficiently compact. Even ordinary binaries supply energy to a cluster as their binding energy grows. The presence of a small number of binaries may inject enough energy to slow or reverse core collapse.

Another oversimplification has been to ignore the finite sizes of the stars. When a cluster contracts sufficiently a significant number of collisions may occur. Their effect is complex, for many processes are possible: partial or complete disruption of the stars, re-accretion of disrupted material by stars, or stellar coalescence. The qualitative effect of these collisions is probably to accelerate the cluster evolution and to produce a small number of more massive stars. These rapidly evolve to supernova explosion or collapse. If these processes are important, a cluster core may ultimately either destroy itself as its stars explode, or leave behind a single black hole. Such processes taking place in the nuclei of galaxies may be the genesis of quasars.

The endpoint of globular cluster evolution remains controversial. There is no clear evidence for any relics of core collapse, suggesting either that the relics are not recognizable as globular clusters, or that they are recognized as globular clusters and are not obviously distinguishable from clusters which have not undergone core collapse. The former possibility seems unlikely, for there should be some halo stars in orbits of high angular momentum. These stars never enter the core, and suffer no dynamical evolution, regardless of what happens in the core. They would remain behind as a dilute globular cluster, with little central condensation. It may be that many of the globular clusters we observe are such relics, and that their masses were once much larger than they are today.

2.5 Nonthermal Particle Acceleration

2.5.1 Spectral Shapes If a population of classical particles (or photons) has relaxed to thermodynamic equilibrium at temperature T, then the probability $n(E)$ that a state of energy E will be occupied is proportional to the Boltzmann factor $\exp(-E/k_BT)$. At high particle density the effects of quantum statistics become significant, and this exponential is replaced by $[\exp(a + E/k_BT) \pm 1]^{-1}$, where $-ak_BT$ is the chemical potential, and $+1$ applies to fermions and -1 to bosons; in this discussion we need consider only the classical Boltzmann factor.

The rate of emission $F(\nu)$ of photons of frequency ν from such an equilibrium gas will be of the form

$$F(\nu) = \int d^3p \, n(E)\mathcal{F}(\nu, E), \tag{2.5.1}$$

where $\mathcal{F}$ is the emission rate per unit frequency interval of photons of frequency ν by a particle of energy E. For emission by interactions between two particles n is the density of particle pairs (analogous to f_2 defined in **2.1**) whose relative motion has kinetic energy E. For continuum emission $\mathcal{F}(\nu, E)$ will be a smoothly varying function of ν, dropping to zero at $h\nu = E$. A very rough approximation to $F(\nu)$ may be obtained by writing

$$\mathcal{F}(\nu, E) \sim \begin{cases} \mathcal{F}_0, & \text{for } h\nu \le E; \\ 0, & \text{for } h\nu > E. \end{cases} \tag{2.5.2}$$

Then, using the nonrelativistic relation between p and E (equilibrium is rarely achieved at relativistic temperatures),

$$\begin{aligned} F(\nu) &\sim \mathcal{F}_0 \int_{E=h\nu}^{\infty} d^3p \, \exp(-E/k_B T) \\ &\sim \mathcal{F}_0 \int_{h\nu}^{\infty} \sqrt{E} \, dE \, \exp(-E/k_B T) \\ &\sim \mathcal{F}_0 \exp(-h\nu/k_B T); \end{aligned} \tag{2.5.3}$$

in the same spirit as (2.5.2) slowly varying factors have been ignored. If $\mathcal{F}(\nu, E)$ consists of narrow spectral lines (2.5.3) is the envelope of a series of narrow spikes corresponding to the spectral lines.

The exponential dependence of $n(E)$ on E and T produces a spectrum $F(\nu) \sim \exp(-h\nu/k_B T)$. If $h\nu \gtrsim k_B T$ this exponential usually gives the dominant frequency dependence of $F(\nu)$ (other than the shapes of individual spectral lines, if present), because it is usually a much more sensitive function of ν than any dependence of $\mathcal{F}(\nu, E)$ on ν which we neglected in (2.5.2). Radiation produced by

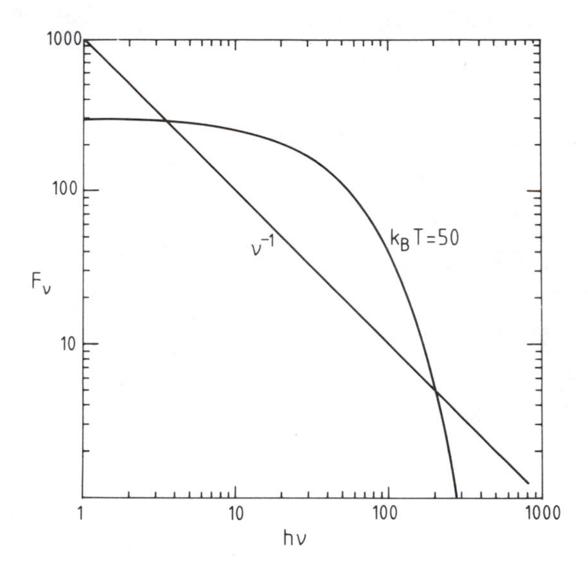

Figure 2.1. Spectral shapes.

matter in thermal equilibrium almost invariably has the exponential frequency dependence of (2.5.3) at frequencies $\nu \gtrsim k_B T/h$. At lower frequencies $F(\nu)$ depends more on the form of $\mathcal{F}(\nu, E)$ and on the optical depth of the source, approaching the Planck function (1.7.13) if the optical depth is high.

When plotted on a log-log plot, (2.5.3) has a characteristic curved shape, and defines the characteristic energy $k_B T$ (Figure 2.1). Many astronomical objects have spectra like this, and it is often possible to recognize the shape (2.5.3) and to estimate $k_B T$ at a glance, even from poor quality data.

Other astronomical objects have spectra which are nearly

straight lines on a log-log plot, corresponding to power law spectra

$$F(\nu) \propto \nu^{-s}, \tag{2.5.4}$$

where the spectral index s is nearly constant. Such a spectrum (and only such a spectrum) defines *no* characteristic frequency, and hence no characteristic energy of the radiating particles. An example is shown in Figure 2.1.

A power law spectrum of the form (2.5.4) cannot be extrapolated indefinitely to both high and low frequency, because if it were the integrated power $\int F(\nu)\, d\nu$ would diverge. There must be at least one change in s, and the frequency ν_0 at which it occurs defines a characteristic energy $h\nu_0$. However, a power law spectrum is often found over several decades of frequency, which indicates that over a large range in energy of the radiating particles their distribution function $n(E)$ also has no characteristic energy. Such a $n(E)$ is also a power law, and may have this form over a very wide range in energy, but similarly must have at least one break in its slope in order that the integrated number density and energy density be finite.

Bodies in thermal equilibrium may produce power law spectra at frequencies for which $h\nu \ll k_B T$. At low energies $n(E)$ is nearly constant (or proportional to E for bosons with zero chemical potential), which are (uninteresting) examples of distributions with no characteristic energy. The simplest and most familiar example of a thermal equilibrium power law spectrum is that of a black body at $h\nu \ll k_B T$, called a Jeans spectrum, for which $s = -2$; the power law slope is broken at $h\nu \approx k_B T$. Optically thin thermal emitters produce power law spectra for $h\nu \ll k_B T$ if their emissivity integrated over the distribution function is a power law function of ν; bremsstrahlung (**2.6.1**) is a familiar example. Thermal emitters with power law distributions of temperature also produce power law spectra, as discussed in **3.6** for accretion discs. In this case the power law slope is broken at $h\nu \approx k_B T_c$, where T_c is a characteristic temperature (typically a break in the distribution of temperature).

The most interesting sources of power law spectra are those in which there is no characteristic energy because the particles are *not* in thermal equilibrium, but instead $n(E)$ is a power law over a wide range of energies. Usually these are very relativistic energies $E \gg mc^2$; the rest mass is another characteristic energy and will generally interrupt both a power law $n(E)$ and the power law spectrum it radiates. It is equally possible to have a power law $n(E)$ for $E \ll mc^2$, but such lower energy nonthermal particles are usually less interesting as sources of astronomical radiation; they are significant in laboratory and astrophysical plasma physics. The presence of power law $n(E)$ are inferred from observations of power law $F(\nu)$ in quasars, active galactic nuclei, extragalactic radio sources, many Galactic radio sources (particularly those associated with supernova remnants), and a variety of compact objects. A power law $n(E)$ is directly observed for cosmic rays, with energies extending to $\gtrsim 10^{19}$ eV.

2.5.2 Particle Acceleration In order to understand the origin of a nonequilibrium particle distribution function $n(E)$ it is necessary to understand the processes by which particles gain and lose energy. In **2.2** and **2.3** we considered the relaxation towards thermal equilibrium of distributions of charged particles and photons in contact with a heat bath. These processes cannot produce power law distributions of very energetic particles, because their energies exceed the thermal energy per particle of any conceivable heat bath. We are here concerned with processes which can accelerate energetic particles as the consequence of many small increments to their energy.

Fermi (1949) suggested a mechanism for the acceleration of energetic cosmic rays (Longair 1981); it may be applied to particles scattering within any dilute moving medium, such as discs (**3.6**) or accretion flows (**3.7**). This mechanism is called stochastic acceleration because it involves numerous uncorrelated random events. Particles are accelerated as a result of their scattering by moving massive objects. In the interstellar medium these objects could be interstellar clouds or magnetohydrodynamic waves; charged particles are elastically scattered by clouds by being deflected by their magnetic fields. This process may be thought of as the relaxation of the particles and the clouds (considered as single very massive particles) towards thermal equilibrium. Because the kinetic energies of the clouds are enormous ($\sim 10^{45}$ erg for plausible interstellar clouds, corresponding to $T \sim 10^{61}$ °K), equilibrium can never be reached, but there will be a net energy flow from the kinetic energy of the clouds to that of relativistic particles. There is, in principle, no characteristic energy scale between the particle rest mass energy mc^2 and the kinetic energy of an entire cloud, so that a power law distribution of particle energy may be achieved over a very large range in energy (in practice, various energy loss mechanisms have thresholds which may serve to define characteristic energies).

A relativistic particle of energy E elastically scattered by an object of velocity $v \ll c$ suffers an energy change

$$\Delta E \sim \mathcal{O}\left(E\frac{v}{c}\right). \tag{2.5.5}$$

This change in energy is as likely to be positive as negative, for isotropically distributed cloud velocities. There will, however, be a mean positive energy shift

$$\langle \Delta E \rangle \sim \mathcal{O}\left(E\frac{v^2}{c^2}\right). \tag{2.5.6}$$

The numerical coefficient in (2.5.6) is of order unity but depends on the detailed geometry and kinematics of the scattering, which depend on the (unknown) details of cloud magnetic field geometry. This energy transfer process is analogous to Comptonization (**2.3**) in the soft photon limit, with clouds instead of electrons and the relativistic particles instead of low energy photons; the rate of energy transfer from the electrons (clouds) to the photons (particles) is proportional to the product of the photon (particle) energy density and the electron (cloud) kinetic energy density (2.3.31).

After $\mathcal{N} \gg 1$ scatterings a relativistic particle whose initial energy was E_0 has an energy of approximately

$$E \approx E_0 \exp(\beta_s^2 \mathcal{N}), \tag{2.5.7}$$

where the parameter $\beta_s \approx v/c$. If $E \gg E_0$ the dispersion in $\ln(E/E_0)$ will be less than its mean value, so it is reasonable to take the mean value and ignore the dispersion, as in (2.5.7).

Particles will not gain energy forever. Eventually losses will become significant. Consider sudden random catastrophic losses, such as escape from the region in which acceleration takes place, and (for strongly interacting particles) nuclear collisions in which most of their energy is lost. It the characteristic loss time is T, independent of energy, then the fraction of particles surviving for times between t and $t + dt$ is

$$N(t)\ dt = \frac{\exp(-t/T)}{T} dt. \tag{2.5.8}$$

If the mean rate of scattering of particles by clouds is $1/\tau$, then $\mathcal{N} \approx t/\tau$ (which holds accurately for $t \gg \tau$), and after a time t a particle has the energy

$$E(t) \approx E_0 \exp(\beta_s^2 t/\tau). \tag{2.5.9}$$

The energy distribution is found

$$\begin{aligned} N(E)\ dE &= N(t) \frac{dt}{dE} dE \\ &\approx \frac{\exp(-t/T)}{T} \frac{\tau}{\beta_s^2 E} dE. \end{aligned} \tag{2.5.10}$$

Using (2.5.9) to express t in terms of E yields

$$N(E) \approx \frac{\tau}{\beta_s^2 T E_0} \left(\frac{E}{E_0} \right)^{-(1+\tau/\beta_s^2 T)} . \qquad (2.5.11)$$

This is a power law distribution, and has the spectral index $p = 1 + \tau/\beta_s^2 T$. $N(E)$ is the total number of particles per unit energy interval; for relativistic particles or photons $N(E) \propto E^2 n(E)$. The absence of a characteristic energy, as expressed in the exponential functions of (2.5.8) and (2.5.9), makes the power law form inevitable. A very similar derivation applies to soft photon Comptonization (**2.3**) in a finite volume, in which escape from the volume is the loss process, and power laws are similarly predicted.

In problems of interest the spectral index p is of order unity; for cosmic rays it is observed to be close to 2.6, and for relativistic electrons in sources of synchrotron radiation it is typically between 2 and 3. There is probably no deep significance to the fact that p is generally found in such a narrow range. For $p \leq 2$ the total energy density diverges as $E \to \infty$, so values smaller than 2 are unlikely to describe $N(E)$ over a very wide range of energy. For $p \gtrsim 3$ the energy contained in very energetic particles is very small; the high energy parts of such steep distribution functions therefore produce little radiation, which is less likely to be detected than that produced by regions or objects in which p is smaller.

There is no obvious *a priori* reason why the parameters τ, β_s, and T should be as closely constrained as is p. For cosmic rays it is known from studies of their nuclear abundances that $T \sim 10^7$ yr, while from observations of the interstellar medium $\beta_s \approx 3 \times 10^{-5}$; τ depends on the microstructure of interstellar magnetic fields and can only be guessed at. For regions in which energetic electrons are accelerated all these parameters are very uncertain. The likely explanation is that the energetics of the accelerating process is self-regulating. The mean energy per particle is $E_0(p-1)/(p-2)$. As

$p \rightarrow 2$ the power required to accelerate the energetic particles diverges, so particle acceleration becomes a strong damper on the cloud motions or waves which drive it. As $p \rightarrow 3$ the power required to accelerate the particles becomes a small multiple of the power required to inject them at energy E_0. β_s and τ (which depend on the detailed distribution of fluid velocity, wave amplitude, and magnetic field) may adjust themselves so that the power supplied to the energetic particles equals that supplied to the accelerating clouds or waves (for cosmic rays in our Galaxy probably the kinetic energy of supernovae). This naturally leads to $2 < p \lesssim 3$ for a wide range of parameters.

We have so far assumed an energy change quadratic in $\beta_s \ll 1$. A particle may occasionally be trapped between two converging reflectors. Then the energy change on each scattering is positive, and the mean energy increment is linear in β_s. In a chaotic flow periods when a particle is trapped between converging reflectors are balanced, in part, by those when it is trapped between diverging reflectors (but not completely balanced, because as reflectors converge the mean time between scatterings becomes progressively smaller, while as they diverge it becomes larger). In an ordered converging flow the particles will always be in a region of convergence, in which case β_s^2 should be replaced by β_s in the previous results. This leads to much more rapid acceleration. An example of such a converging flow is that around a shock (**3.3**); if the fluid on each side contains magnetic disturbances capable of scattering the particles they will gain energy rapidly. Shocks are therefore promising locations for particle acceleration. It may even be that much of the kinetic energy released in some shocks appears in accelerated high energy particles rather than the internal energy of the shocked fluid.

Fermi acceleration is capable of accelerating high energy particles to yet higher energy, but does not answer the question of their initial acceleration. The theory is applicable to nonrelativistic particles, if their speed v_p is used in place of c, but at low energy their rate of slowing by Coulomb drag (**2.2**) exceeds their rate of acceleration. Approximately, the loss rates by Coulomb drag, hard nuclear collisions, and escape may be added to give T^{-1}; if $T \sim \tau/\beta_s^2$ at relativistic energies, as must be the case, then for $E \ll mc^2$ Coulomb drag will probably overwhelm any acceleration. However, it is conceivable that if there are no relativistic particles τ/β_s^2 will decrease until it equals the Coulomb slowing time for a few of the fastest thermal particles; once these particles are accelerated to relativistic energy it will increase again as they rapidly gain energy and the energy balance regulates itself. If this does not occur (and there is no evidence for it), there is a large gap in energy between thermal particles and those moving fast enough for Fermi acceleration to be effective.

There is another process, analogous to Fermi acceleration but involving plasma oscillations instead of magnetized gas clouds or hydromagnetic waves, which may be capable of bridging the energy gap. This process draws energy from the plasma waves and is called Landau damping, but because it accelerates particles it may also be called Landau acceleration. Its essential property is that it is resonant, meaning that a given plasma wave can accelerate only particles with a very narrow range of velocity. As a result, it may be effective in accelerating low energy particles, where rapid acceleration is necessary to overcome Coulomb drag, without an excessive amount of energy being drained from the plasma waves by the further acceleration of particles which are already relativistic. The theory of Landau damping is subtle, but is explained in innumerable books on plasma physics; see, for example, Nicholson (1983). Here we present only a simple and elementary argument.

Consider a particle of charge e and velocity v suddenly placed in an electric field

$$E(x,t) = E_0 \cos(kx - \omega t). \tag{2.5.12}$$

We ignore the subtlety of the problem by replacing a careful investigation of the initial conditions with a "suddenly." The particle is accelerated

$$m\frac{dv}{dt} = eE_0 \cos(kx - \omega t). \tag{2.5.13}$$

If E is in the x-direction (a longitudinal wave) $x \sim \int v\, dt$, and it is necessary to expand v in a series in powers of $eE_0/m\omega$, keeping terms of the second order. A similar result is obtained much more simply for transverse waves. In this case we can write $x = v_x t$ and let (2.5.13) describe the particle motion in the direction of the electric field, taken to be z. Then

$$v_z = \frac{eE_0}{m(kv_x - \omega)} \sin[(kv_x - \omega)t], \tag{2.5.14}$$

and the particle's transverse motion has a kinetic energy

$$\frac{1}{2}mv_z^2 = \frac{e^2E_0^2}{2m(kv_x - \omega)^2} \sin^2[(kv_x - \omega)t]. \tag{2.5.15}$$

The resonant nature of the acceleration is evident; particles whose velocity v_x is close to the wave phase velocity ω/k receive a great deal of energy. We assume nonrelativistic motion, and integrate (2.5.15) over the velocity distribution function $f(v_x)$, taken to be slowly varying near $v_x = \omega/k$. Then the total kinetic energy imparted to the particles by the wave is

$$\begin{aligned} K &= \int \frac{1}{2}mv_z^2 f(v_x)\, dv_x \\ &\approx \frac{e^2E_0^2 f(\omega/k)}{2m} \int \frac{\sin^2[(kv_x - \omega)t]}{(kv_x - \omega)^2} dv_x \\ &\approx \frac{e^2E_0^2 f(\omega/k)t}{2mk} \int_{-\infty}^{\infty} \frac{\sin^2 u}{u^2} du \\ &\approx \frac{\pi e^2E_0^2 f(\omega/k)t}{2mk}, \end{aligned} \tag{2.5.16}$$

and the power is

$$P \approx \frac{\pi e^2 E_0^2 f(\omega/k)}{2mk}. \tag{2.5.17}$$

The approximations in these equations approach equality as $t \to \infty$ and the integrand becomes sharply peaked.

The calculation for longitudinal waves (Nicholson 1983) shows that particles with $v_x < \omega/k$ gain energy from the wave, while those with $v_x > \omega/k$ give up energy to it. The total power supplied to the particles resembles (2.5.17), but in place of $f(\omega/k)$ there is $(\omega/k)f'(\omega/k)$ (which is generally of the same order of magnitude).

Particles moving with the phase velocity of plasma waves will be accelerated by the waves. Because this acceleration is random in direction and sign (depending on phase and polarization for transverse waves, and phase and the sign of $\omega/k - v_x$ for longitudinal waves), particles diffuse in momentum space. This resembles the **b** term in the Fokker-Planck equation (2.1.10), the $\partial^2 n/\partial x^2$ term in the Kompaneets equation (2.3.21), and the scattering of relativistic particles by moving magnetized clouds or hydromagnetic waves (2.5.7). If the plasma waves are only excited to a thermal level (an average of $\frac{1}{2}k_B T$ per degree of freedom) this diffusion is only a small increment to the diffusion (2.2.17) produced by encounters between the particles, and is exactly cancelled by an additional drag term resulting from the emission of plasma waves by particles. However, if plasma waves are excited to a high intensity by a plasma instability, the rate of momentum diffusion may be large and the rate of acceleration rapid. A variety of plasma waves exist with a broad range of phase velocities and may be excited to high intensity, so it may be possible by this mechanism to accelerate particles from the high velocity tail of a Maxwellian distribution to much higher (possibly even relativistic) velocities. This phenomenon is observed in many laboratory plasma experiments, and is a frequent consequence of plasma instability.

It is also possible to accelerate energetic particles with a single large potential drop, without the necessity of their diffusing in energy. This is believed to occur in pulsars (**4.4**). It may not occur elsewhere, because astrophysical plasmas are usually good conductors, and it is not easy to produce large potential drops, although they may occur transiently as a consequence of complex magnetohydrodynamic flows called "field reconnection," in which current densities are high and scattering by plasma waves impedes the flow of current and leads to a temporarily large resistivity. Acceleration by a single potential drop does not lead to the observed power law particle distribution functions, but the effects of many potential drops, with a power law distribution of accelerating potentials, may.

Colgate and Johnson (1960) suggested that purely hydrodynamic processes could accelerate power law distributions of energetic particles. As a shock travels into a medium of progressively decreasing density (the atmosphere of an exploding star **4.3**, or an interstellar density gradient) the velocity and energy (per particle) of the shocked fluid progressively increase. Although locally it may be in thermal equilibrium, the overall distribution of kinetic energy per particle approximates a power law. This mechanism has difficulty explaining some of the observed properties of cosmic rays, for which it was proposed, but it may be capable of providing the initial acceleration across the velocity range in which Coulomb drag is important.

A great wealth of detailed mechanisms for particle acceleration have been discussed (Arons, *et al.* 1979); it is likely that different ones are important in different objects. All of them depend on the detailed distribution of fluid velocity, magnetic field, or plasma waves. For this reason acceleration is difficult to calculate even in laboratory and Solar System plasmas, where direct probes and abundant data exist. In a remote astronomical object the problem is much harder.

Phenomena involving nonthermal particles pose the most difficult problems in astrophysics. They are nearly ubiquitous, and are observed in the solar photosphere as a consequence of turbulent convection, in planetary magnetospheres as a consequence of interaction with the Solar wind, as well as in more exotic and distant places. Much of the power of some pulsars, quasars, and γ-ray sources appears as energetic particles, but there is no quantitative understanding or theory. Although it is not possible to predict the properties of nonthermal phenomena, the astrophysicist should not be surprised by their appearance. They are likely to be found wherever fluid motions are available as a source of free energy, whether turbulent or ordered, as in accretion, shocks, or the differential rotation of a disc.

2.6 Radiation Processes

Matter absorbs, emits, and scatters radiation by a variety of processes. These are usually the chief mechanisms by which radiation and matter relax towards thermal equilibrium. Nearly all our knowledge of astronomical objects is obtained from the study of their radiation, and their properties and structure usually are determined by the emission and absorption of radiation within them. Fortunately, most radiation processes are well understood. Their calculation is often tedious or difficult, but quantitative results are available. In this section I present a few simple results for important processes. I have drawn upon the book of Rybicki and Lightman (1979), which contains more complete discussions and derivations of most of these results, and references to the literature.

2.6.1 Bremsstrahlung When a charged particle is accelerated it radiates. An important example is the acceleration of an electron by a Coulomb field, in which case the radiation is called bremsstrahlung. In the nonrelativistic limit the instantaneous power radiated by a system of charges is

$$P = \frac{2\ddot{d}^2}{3c^3}, \tag{2.6.1}$$

where d is any one component of the electric dipole moment $\vec{d}$ of the system; contributions from each component are added. In this limit the acceleration of an electron by another electron produces no radiation, because the opposite accelerations of the two electrons produce no net change in $\vec{d}$ (At relativistic energies electron-electron bremsstrahlung is comparable to electron-ion bremsstrahlung). The spectral density radiated in a single encounter is

$$\frac{dW}{d\omega} = \frac{8\pi\omega^4}{3c^3}\left|\hat{d}(\omega)\right|^2, \tag{2.6.2}$$

where $\hat{d}$ is the Fourier transform of d:

$$\hat{d}(\omega) = \frac{1}{2\pi}\int_{-\infty}^{\infty} e^{i\omega t} d(t)\, dt. \tag{2.6.3}$$

A quantitative calculation of bremsstrahlung is lengthy (and must be performed quantum mechanically). Instead, we use these classical expressions and make rough approximations in order to provide an illustriative guide. The dipole moment $\vec{d}$ represented by an accelerated electron of charge $-e$ at position $\vec{r}$ is

$$\vec{d} = -e\vec{r}. \tag{2.6.4}$$

The fixed origin of coordinates is irrelevant, because we are only interested in time derivatives of d, but may be conveniently taken at the ion of charge Ze from which the electron scatters, or the center of mass of the two particles. Taking two time derivatives of (2.6.4) and considering only one component gives

$$\ddot{d} = -e\dot{v}. \tag{2.6.5}$$

The Fourier transform of (2.6.5) is

$$\int e^{i\omega t}\ddot{d}\, dt = -e \int e^{i\omega t}\dot{v}\, dt. \tag{2.6.6}$$

Integrating the left hand side by parts twice, and using (2.6.3), yields

$$\omega^2 \hat{d}(\omega) = \frac{e}{2\pi} \int e^{i\omega t}\dot{v}\, dt. \tag{2.6.7}$$

Now consider an electron approaching an ion at speed v and impact parameter b. If its deflection is not large we may calculate the force on and acceleration of the electron as if its path were a straight line. Most (55%) of its radiation occurs during a time interval $\tau \equiv b/v$ when the electron is within a distance $b\sqrt{5/4}$ of the ion; outside this interval the power is less than 0.64 of its peak value. The cumulative impulse and change in velocity Δv is found by integrating along the undeflected trajectory the component of acceleration normal to it:

$$\begin{aligned} \Delta v &= \frac{-Ze^2}{m_e} \int_{-\infty}^{\infty} \frac{b\, dt}{(b^2 + v^2t^2)^{3/2}} \\ &= -\frac{2Ze^2}{m_e bv}; \end{aligned} \tag{2.6.8}$$

the impulse parallel to the trajectory integrates to zero in this approximation.

For very high frequencies $\omega\tau \gg 1$ and the exponential in (2.6.7) has many cycles of oscillation over the range over which $\dot{v}$ varies significantly, so the positive and negative phases of the integrand nearly cancel. At low frequencies $\omega\tau \ll 1$ and the exponential may be taken to be unity where $\dot{v}$ is significantly different from zero. Considering only these two limits, and ignoring the more difficult intermediate regime, gives

$$\hat{d}(\omega) \approx \begin{cases} \dfrac{e\Delta v}{2\pi\omega^2} & \text{for } \omega\tau \ll 1; \\ 0 & \text{for } \omega\tau \gg 1. \end{cases} \tag{2.6.9}$$

Substitution into (2.6.2) gives

$$\frac{dW}{d\omega} \approx \begin{cases} \dfrac{8Z^2e^6}{3\pi c^3 m_e^2 b^2 v^2} & \text{for } b \ll v/\omega; \\ 0 & \text{for } b \gg v/\omega. \end{cases} \tag{2.6.10}$$

It is now necessary to integrate over the distribution of inpact parameters to obtain the spectral density $dP/d\omega$ radiated by an electron with speed v passing through a gas of ions of density n_i:

$$\begin{aligned} \frac{dP}{d\omega} &= n_i v \int_0^\infty \frac{dW}{d\omega} 2\pi b \, db \\ &\approx \frac{16 n_i Z^2 e^6}{3c^3 m_e^2 v} \int_{b_{min}}^{v/\omega} \frac{db}{b}. \end{aligned} \tag{2.6.11}$$

A lower cutoff b_{min} has been introduced into the integration to avoid the logarithmic divergence at $b \to 0$. Physically, the origin of b_{min} is the breakdown of either the small deflection approximation for $b \lesssim Ze^2/m_e v^2$ or of the classical approximation for $b \lesssim h/m_e v$. A more quantitative calculation must include both these effects, and is rather lengthy. The ratio between the accurate result and a rough approximation like ours is called the Gaunt factor g. This factor has been calculated in detail, and is usually of order unity. The spectral density is

$$\frac{dP}{d\omega} = \frac{16\pi n_i Z^2 e^6}{3\sqrt{3} m_e^2 v} g(v,\omega), \tag{2.6.12}$$

where the extra factor of $\pi/\sqrt{3}$ is required by the standard definition of g.

To compute the total power radiated integrate (2.6.12) over frequency, cutting off the integration at $\hbar\omega = \frac{1}{2}mv^2$ because photon energy is quantized (without this cutoff, which is implicit in g, the power would diverge). The result is

$$P_{brems} = \frac{8\pi n_i Z^2 e^6 v \langle g \rangle}{3\sqrt{3} c^3 m_e \hbar}. \tag{2.6.13}$$

This power may be compared to the rate at which the electron loses energy to other electrons by Coulomb drag (using 2.2.9; $n_e = Zn_i$ and $m_{12} = m_e/2$). Their ratio is

$$\frac{P_{brems}}{P_{drag}} = \frac{Z\langle g \rangle}{3\sqrt{3}\ln\Lambda} \frac{v^2}{c^2} \frac{e^2}{\hbar c}. \tag{2.6.14}$$

The last factor is the fine structure constant α, and is nearly equal to 1/137. It is evident that for nonrelativistic electrons (the only case to which these results are applicable), bremsstrahlung energy loss is very small compared to Coulomb drag.

Now integrate (2.6.12) over a Maxwellian distribution of electron velocities. To do this quantitatively requires knowledge of $g(v, \omega)$; in the result this is absorbed into a new integrated Gaunt factor $g(T, \omega)$. The integration begins at $v = \sqrt{2\hbar\omega/m}$, because lower speed electrons cannot produce a photon of frequency ω. The resulting emissivity is

$$j(T, \omega) = 7 \times 10^{-38} \frac{n_e n_i Z^2}{T^{1/2}} \exp(-\hbar\omega/k_B T) g(T, \omega) \frac{\text{erg}}{\text{cm}^3 \text{ sec Hz}}. \tag{2.6.15}$$

The physical origin of the $T^{-1/2}$ factor is the v^{-1} in (2.6.12), which in turn comes from the square of the v^{-1} in (2.6.8) and the v in (2.6.11). The characteristic exponential comes from the Maxwellian distribution function, as discussed in **2.5.1**. Because this contains the only strong dependence on ω in (2.6.15), the spectrum of thermal bremsstrahlung resembles the exponential shown in figure 2.1.

Such a spectrum is actually observed from dilute clouds of ionized interstellar gas. Bremsstrahlung is believed to be important in many more compact objects, but in these objects the optical depth (1.7.23) is large, re-absorption of the radiation is important, and the emergent spectrum (**1.14**) does not resemble the source function (2.6.15); in the limit of large optical depth (for example, a star) the emergent spectrum is close to a Planck function (1.7.13).

2.6.2 Magnetic Radiation An electron moving across a magnetic field $\vec{B}$ is accelerated by the field, and therefore radiates. This is known as cyclotron emission. The electron follows a helical path, gyrating about $\vec{B}$ with an angular frequency

$$\omega_B = \frac{eB}{\gamma m_e c}, \tag{2.6.16}$$

where $\gamma \equiv (1 - v^2/c^2)^{-1/2}$, and a radius of gyration

$$r_g = \frac{\gamma m_e c v_\perp}{eB}, \tag{2.6.17}$$

where $v_\perp$ is the magnitude of the component of velocity perpendicular to $\vec{B}$.

For nonrelativistic motion ($\gamma \to 1$) the radiated power is found from (2.6.1). There are two oscillating components of $\vec{d}$, each varying sinusoidally with amplitude er_g and frequency ω_B, but 90° out of phase. The total radiated power is then

$$\begin{aligned} P_{cyc} &= \frac{2\omega_B^4 e^2 r_g^2}{3c^3} \\ &= \frac{2e^4 B^2 v_\perp^2}{3m_e^2 c^5}. \end{aligned} \tag{2.6.18}$$

Cyclotron emission by ions is generally negligible because of the two powers of m in the denominator.

Because the motion of a gyrating electron is sinusoidal in time, in the nonrelativistic limit the radiation is monochromatic at the frequency $\nu_B = \omega_B/2\pi$. The component $v_{\parallel}$ of electron velocity parallel to $\vec{B}$ leads to significant Doppler broadening of the radiation received by all observers except those in directions exactly perpendicular to $\vec{B}$. In a real astronomical object the magnetic fields are curved, and all observers will receive Doppler-broadened radiation; B and ν_B will also vary from place to place within the emission region.

The kinetic energy $\frac{1}{2}m_e v_{\perp}^2$ of perpendicular motion is reduced by cyclotron radiation, and is easily seen to decay exponentially with an e-folding time

$$\begin{aligned} t_{cyc} &= \frac{3m_e^3 c^5}{4e^4 B^2} \\ &= 2.6 \times 10^{-4} \left(\frac{B}{10^6 \text{ gauss}}\right)^{-2} \text{ sec.} \end{aligned} \tag{2.6.19}$$

This time is short for large fields, such as those found in magnetic white dwarves ($B \sim 10^7$ gauss) and magnetic neutron stars ($B \sim 10^{12}$ gauss), and implies that electrons rapidly radiate the kinetic energy of their perpendicular motion. In such large fields cyclotron radiation is usually the most rapid radiation process. If t_{cyc} is small compared to t_{eq} (2.2.24), and the cyclotron radiation freely escapes, the electron distribution function will become strongly anisotropic.

Cyclotron radiation is, in general, elliptically polarized. This is a general property of radiation processes in strong magnetic fields, and is unusual in astrophysics. The elliptical polarization of a few white dwarves led to the recognition that they have large magnetic fields, although cyclotron emission is not usually the dominant source of radiation.

Because electrons in large magnetic fields are such efficient radiators, they are also efficient absorbers of radiation at their cyclotron frequency. It is generally incorrect to apply (2.6.18) or (2.6.19) directly to the radiation of a gas of electrons, because the radiation emitted by one will be efficiently absorbed by its neighbors. In order to estimate the opacity we need to assume a finite line width $\Delta\nu$. The Doppler width will typically range from $\sim .001\nu$ for cool white dwarves to $\sim .1\nu$ for hot accreting neutron stars; the cyclotron line will also be broadened by a variety of collisional and plasma processes. From (2.6.18) and (1.7.13) we obtain the cyclotron absorption opacity

$$\begin{aligned}\kappa_{cyc} &= \frac{8\pi^3}{3}\left(\frac{\nu}{\Delta\nu}\right)\frac{k_B T m_e c}{B^2 h m_p \mu_e}\Big(1 - \exp(-h\nu_B/k_B T)\Big) \\ &\approx 3 \times 10^{11}\left(\frac{.01\nu}{\Delta\nu}\right)\left(\frac{B}{10^6 \text{ gauss}}\right)^{-1} \text{cm}^2/\text{gm},\end{aligned} \tag{2.6.20}$$

where μ_e is the molecular weight per electron. In the numerical evaluation $\mu_e = 1.2$ (ordinary stellar matter) and $h\nu_B \ll k_B T$ were assumed. The corresponding cross-section is $\sim e^2/(\Delta\nu m_e c)$, which is the natural cross-section for absorption by an electron considered as a classical oscillator.

This is usually a very large opacity, much exceeding that produced by other processes, but applicable only within the spectral width $\Delta\nu$ of the line. Cyclotron line radiation therefore flows very slowly through a magnetized plasma, and the rate at which it radiates is usually much less than (2.6.18) or (2.6.19) would imply. For a thermal electron distribution function the spectral power density produced by cyclotron emission cannot exceed the Planck function (1.7.13). At frequencies far from the line center there is no emission at all. When the optical depth at the line center is high, the width of the line becomes important, and broadening processes must be considered.

Because of special relativistic kinematics the motion of an electron seen by an unaccelerated observer is not exactly harmonic. Consequently, the radiation is not strictly harmonic at the frequency ν_B. Because the motion is still periodic (if radiation damping and collisions are ignored) the spectrum is a series of harmonics of ν_B, with the strength of the n-th harmonic varying $\sim (v_\perp^2/c^2)^n$. The dependence on electron energy and frequency $n\nu_B$ is steep, and cyclotron harmonic radiation therefore does not follow (2.5.3).

The importance of the harmonics is that the spectral density may approach the Planck function B_ν in each of them. If $\Delta\nu/\nu$ is independent of n and $nh\nu_B \ll k_B T$, then the power in the n-th harmonic is $\sim \Delta\nu B_\nu \sim \nu^3 \sim n^3$, while the total power in harmonics $1 - n$ is $\sim n^4$. Because radiation at the fundamental frequency is so strong, rather high harmonics may be produced, especially at the high temperatures characteristic of accreting degenerate dwarves, neutron stars, and laboratory plasma machines; harmonics up to $n \sim 10-100$ may be optically thick. Under these conditions the harmonic lines usually overlap because of Doppler broadening to form a smooth continuum, $\Delta\nu$ is effectively constant, and the total radiated power is $\sim n^3$. Quantitative calculations are intricate; Petrosian (1981) gives some results and references to the earlier literature.

The radiation produced by a relativistic electron in a magnetic field, called synchrotron radiation, is also of interest. Most inferences of the presence and acceleration of energetic electrons are based on the observation of their synchrotron radiation, usually at radio frequencies (in supernova remnants and extragalactic radio sources) or in visible light (in many quasars and active galactic nuclei, and the Crab nebula), but occasionally in X-rays (in the Crab nebula, and probably other objects). This radiation is readily identified because it is strongly linearly polarized, and has a featureless power law spectrum (when produced by electrons with a power law distribution of energies); there is usually no plausible alternative way of producing radiation with these properties.

The relativistic generalization of (2.6.18) is (Rybicki and Lightman 1979, Jackson 1975)

$$P_{synch} = \frac{2\gamma^2 e^4 B^2 v_\perp^2}{3m_e^2 c^5}. \tag{2.6.21}$$

Most of the emitted radiation is at frequencies a few times lower than a characteristic frequency (defined differently by different authors)

$$\begin{aligned} \nu_c &\equiv 3\gamma^3 \nu_B \sin\alpha \\ &= \frac{3\gamma^2 eB \sin\alpha}{2\pi m_e c}, \end{aligned} \tag{2.6.22}$$

where α is the pitch angle of the electron's helical motion ($\alpha = 0$ for motion parallel to $\vec{B}$, and $\alpha = \pi/2$ for circular motion in a plane normal to $\vec{B}$). This corresponds to harmonic numbers $n \sim 3\gamma^3 \sin\alpha$; for $\gamma \gtrsim 1$ the harmonics overlap and the spectrum is a smoothly varying continuum.

The integrated spectrum produced by electrons with a power law distribution of energies

$$N(E) = N_0 E^{-p} \tag{2.6.23}$$

is given by (2.5.1). Using $E = \gamma m_e c^2$ and $n(E)\,d^3p = N(E)\,dE$, and taking $\gamma \gg 1$, we obtain

$$F(\nu) \sim \int N_0 \gamma^{-p} \mathcal{F}(\nu, \gamma)\, d\gamma. \tag{2.6.24}$$

Now

$$\mathcal{F}(\nu, \gamma) \sim \frac{P_{synch}}{\nu_c} \mathcal{S}(\nu/\nu_c), \tag{2.6.25}$$

where the single function $\mathcal{S}$ describes the shape of the emission spectrum as a function of ν/ν_c at all relativistic energies. Then, using (2.6.21) and (2.6.22), and noting that ν is a variable independent of

γ or ν_c,

$$F(\nu) \sim \int \gamma^{-p} S(\nu/\nu_c)\, d\gamma$$
$$\sim \int \nu_c^{-p/2} S(\nu/\nu_c) \nu_c^{3/2}\, d(\nu_c^{-1}) \tag{2.6.26}$$
$$\sim \nu^{-(p-1)/2} \int (\nu/\nu_c)^{(p-3)/2} S(\nu/\nu_c)\, d(\nu/\nu_c).$$

The last integral depends only on the function S and the limits of integration; if the form (2.6.23) extends over a wide range in E then over a wide range of ν these limits may be taken to be 0 and ∞, and the integral is a number independent of ν. The integrated spectrum is then a power law (2.5.4) with spectral index

$$s = \frac{p-1}{2}. \tag{2.6.27}$$

This power law and linear polarization are the characteristic signatures of a synchrotron source; most frequently $0.5 < s < 1.0$, corresponding to $2 < p < 3$. In visible and ultraviolet light such a spectrum is readily distinguishable from stellar spectra, even by a crude comparison of the colors measured through broad filters. Stellar spectra have a pronounced thermal curvature (Figure 2.1), and therefore stars are almost always fainter in the ultraviolet than power law sources with similar colors in visible light. This permits the quick identification of candidate nonthermal sources (such as quasars) from crude photographic measures of their colors.

The synchrotron power radiated per unit volume is $\sim N_0\gamma^2B^2$. For radiation of a given frequency ν, γ and B are related by (2.6.22); eliminating γ, the power is $\sim \mathcal{E}_e\mathcal{E}_B^{3/4}$, where $\mathcal{E}_e \sim N_0\gamma$ and $\mathcal{E}_B \sim B^2$ are respectively the relativistic electron and magnetic energy densities. For a given total energy density $\mathcal{E} = \mathcal{E}_e + \mathcal{E}_B$ the maximum power is obtained if $\mathcal{E}_e = \frac{4}{3}\mathcal{E}_B$, close to the "equipartition" condition $\mathcal{E}_e = \mathcal{E}_B$. Sources in which equipartition holds, at least roughly, are more efficient radiators and more likely to be observed than those in which it does not hold. Astronomers frequently assume equipartition in order to estimate source parameters; these parameters describe a typical source if (and only if) *some* sources are close to equipartition.

2.6.3 Compton Scattering An electron in an electromagnetic wave will be accelerated by the wave's electric field (Unless the fields are extremely strong, the motion is nonrelativistic and the effect of the magnetic field may be ignored). Such an accelerated electron will radiate. If $\hbar\omega \ll m_e c^2$ a classical description is adequate. The electron's motion is given by (2.5.14). Integrate this expression with $v_x = 0$, and use (2.6.4) to obtain

$$d(t) = \frac{e^2 E_0}{m_e \omega^2} \cos(\omega t). \tag{2.6.28}$$

Using (2.6.1) and integrating over time leads to a mean radiated power

$$\langle P \rangle = \frac{1}{3} \frac{e^4 E_0^2}{m_e^2 c^3}. \tag{2.6.29}$$

This power is drawn from the power of the incident electromagnetic wave, is at the same frequency, and may be described as its scattering by the electron, called Thomson scattering. The mean power density of the electromagnetic wave (including both $\vec{E}$ and $\vec{B}$, and averaging over phase) is $E_0^2 c/8\pi$. The ratio of $\langle P \rangle$ to this power density is the electron scattering cross-section

$$\sigma_{es} = \frac{8\pi}{3} \frac{e^4}{m_e^2 c^4}. \tag{2.6.30}$$

In order to calculate the angular dependence of electron scattering it is necessary to consider the angular dependence of the electron's radiated field. The result (Rybicki and Lightman 1979) for the differential scattering cross-section is

$$\frac{d\sigma_{es}}{d\Omega} = \frac{1}{2}\frac{e^4}{m_e^2 c^4}(1 + \cos^2\theta), \tag{2.6.31}$$

where θ is the angle between the incident and scattered photon directions. Because (2.6.31) is reflection-symmetric about $\theta = \pi/2$ the scattered radiation carries no momentum in the nonrelativistic limit; for most purposes (including Comptonization **2.3**) correct results would be obtained even if electron scattering were taken to be isotropic.

From (2.6.30) the electron scattering opacity is

$$\begin{aligned}\kappa_{es} &= \frac{\sigma_{es}}{\mu_e m_p} \\ &= .20(1 + X)\ \mathrm{cm}^2/\mathrm{gm},\end{aligned} \tag{2.6.32}$$

where μ_e is the molecular weight (in atomic mass units) per electron and X is the mass fraction of hydrogen in the matter. For ordinary stellar composition $X = 0.7$ and $\kappa_{es} = .34\ \mathrm{cm}^2/\mathrm{gm}$.

It is also necessary to consider scattering by electrons moving at relativistic speeds. The photon frequency ν' in the electron's frame is related to its frequency ν in the laboratory frame by the Lorentz transformation

$$\nu' = \nu\gamma(1 - \beta\cos\vartheta), \tag{2.6.33}$$

where v is the electron velocity, $\beta \equiv v/c$, $\gamma \equiv (1 - \beta^2)^{-1/2}$, and ϑ is the angle (in the laboratory frame) between the unscattered photon and electron directions. If $h\nu' \ll m_e c^2$ the scattering may be described as Thomson scattering in the electron's frame. In that frame the frequency shift on scattering is small and may be neglected, as may be the recoil velocity of the electron. The frequency of the scattered photon in the laboratory frame is then

$$\nu'' = \nu'\gamma(1 + \beta\cos\vartheta''), \tag{2.6.34}$$

where ϑ'' is the angle (in the electron's frame) between its velocity and the the scattered photon's direction. The angle ϑ is typically $\sim \pi/2$, if the photon and electron distributions are initially isotropic, and ϑ'' is also typically $\sim \pi/2$ because in the electron's frame the scattering follows the Thomson law (2.6.31); neither of these angles is affected by relativistic beaming. Therefore the factors in parentheses in (2.6.33) and (2.6.34) are generally of order unity and

$$\nu'' \sim \gamma^2 \nu. \tag{2.6.35}$$

For relativistic electrons the photon frequency in the laboratory frame is multiplied by a very large factor. Scattering is roughly equivalent to reflection by a mirror moving at the relativistic electron's speed. Typically, the scattered photon will be an X-ray or γ-ray, even if the unscattered photon was visible light or from the 3 °K background radiation. Because of relativistic beaming nearly all the scattered photons travel in nearly the same direction as the electrons from which they scattered; photons scattered by an isotropic electron distribution are also isotropic.

The result of a more quantitative calculation (Rybicki and Lightman 1979) is that an electron loses energy at the rate

$$P_{Compt} = \frac{4}{3}\sigma_{es} c \gamma^2 \beta^2 \mathcal{E}_r, \tag{2.6.36}$$

where $\mathcal{E}_r$ is the radiation energy density. This may be compared to the energy loss by the synchrotron process (2.6.21), assuming an isotropic electron distribution function so that $\langle v_\perp^2 \rangle = 2\beta^2 c^2/3$, and using (2.6.30):

$$\frac{P_{synch}}{P_{Compt}} = \frac{B^2/8\pi}{\mathcal{E}_r}. \tag{2.6.37}$$

The powers are in the same ratio as the magnetic to the photon energy density, and are often comparable. Synchrotron radiation is observed much more often because it is usually emitted at radio frequencies, where detectors are very sensitive. P_{synch} equals P_{Compt} produced by scattering the 3 °K background radiation if $B \approx 3 \times 10^{-6}$ gauss, a typical interstellar field.

The result (2.6.37) should not be a surprise, because both processes involve the scattering of electromagnetic energy by relativistic electrons. Each process multiplies photon frequencies by a factor $\sim \gamma^2$; in the case of synchrotron radiation the photons are not real, but are effectively present in the acceleration of the electron gyrating around the magnetic field. Relativistic Compton scattering also produces power law photon spectra from power law electron energy distributions. There is usually little danger of confusing these with synchrotron power law spectra because Compton scattered radiation is generally at much higher frequency.

These results for relativistic Compton scattering only apply if $h\nu'' \ll \gamma m_e c^2$ (or, equivalently, $h\nu' \ll m_e c^2$). If these conditions are not met, then the relativistic Klein-Nishina formula must be used for the differential scattering cross-section. Even more important, simple conservation of energy limits the scattered photon energy to $(\gamma - 1)m_e c^2 + h\nu$, and (2.6.35) no longer holds.

2.7 References

Arons, J., McKee, C., and Max, C., eds. 1979, *Particle Acceleration Mechanisms in Astrophysics* (New York: American Institute of Physics).

Chapman, S., and Cowling, T. G. 1960, *The Mathematical Theory of Non-Uniform Gases* 2nd ed. (Cambridge: Cambridge University Press).

Colgate, S. A., and Johnson M. H. 1960, *Phys. Rev. Lett.* **5**, 235.

Eisberg, R. M. 1961, *Fundamentals of Modern Physics* (New York: Wiley).

Fermi, E. 1949, *Phys. Rev.* **75**, 1169.

Grindlay, J. E., Hertz, P., Steiner, J. E., Murray, S. S., and Lightman, A. P. 1984, *Ap. J. (Lett.)* **282**, L13.

Jackson, J. D. 1975, *Classical Electrodynamics* 2nd ed. (New York: Wiley).

Kompaneets, A. S. 1957, *Sov. Phys.—JETP* **4**, 730.

Lewin, W. H. G., and Joss, P. C. 1983, in *Accretion Driven Stellar X-Ray Sources*, eds. W. H. G. Lewin and E. P. J. van den Heuvel (Cambridge: Cambridge University Press), p. 41.

Liboff, R. L. 1969, *Introduction to the Theory of Kinetic Equations* (New York: Wiley).

Light, E. S., Danielson, R. E., and Schwarzschild, M. 1974, *Ap. J.* **194**, 257.

Lightman, A. P., and Shapiro, S. L. 1978, *Rev. Mod. Phys.* **50**, 437.

Longair, M. S. 1981, *High Energy Astrophysics* (Cambridge: Cambridge University Press).

Montgomery, D. C., and Tidman, D. A. 1964, *Plasma Kinetic Theory* (New York: McGraw-Hill).

Nicholson, D. R. 1983, *Introduction to Plasma Theory* (New York: Wiley).

Petrosian, V. 1981, *Ap. J.* **251**, 727.

Rosenbluth, M. N., MacDonald, W. M., and Judd, D. L. 1957, Phys. Rev. **107**, 1.

Rybicki, G. B., and Lightman, A. P. 1979, *Radiative Processes in Astrophysics* (New York: Wiley).

Spitzer, L. 1962 *Physics of Fully Ionized Gases* 2nd ed. (New York: Interscience).

Trubnikov, B. A. 1965, *Reviews of Plasma Physics* **1**, 105.

Chapter 3

Hydrodynamics

3.1 Equations

Most astronomical objects are fluid. In fact, those which are not are exceptional—portions of the terrestrial planets, various small bodies in the solar system, interstellar grains, neutron stars, and perhaps the cores of the oldest white dwarves. Just about everything else the astronomer deals with is fluid—stars, the interstellar medium, flows of gas away from stars, onto stars, and between stars, the clouds which will become stars, and the debris left over after the deaths of stars. Even the stars themselves may be sometimes regarded as the particles making up a fluid. The equations of hydrodynamics, sound, and shocks are discussed in many texts; one of the most elegant is that by Landau and Lifshitz (1959).

Some of the fluids an astronomer deals with are far from thermodynamic equilibrium, and relaxation to equilibrium is very slow; good examples of this are the relativistic particles in the cosmic rays and in nonthermal radio sources, and the stars in a galaxy. But in many cases, particularly for ionized plasmas, local thermodynamic equilibrium applies. This means that at any point in the fluid it may be completely described by a single density, velocity, and temperature. The velocity distribution function of the individual particles is Maxwellian (if they are nondegenerate), and the distribution of their ionization and excitation states is described by the Saha equation. Any processes of excitation, dissociation, ionization, or chemical reaction are either negligibly slow or so rapid that they may be assumed to be in equilibrium. Then, for many purposes, the fact that the particles are charged is inessential, and the ordinary equations of hydrodynamics may be used. We need to remember that we are dealing with a plasma when considering deviations from equilibrium—transport processes, acceleration of particles to high energy, and macroscopic currents and magnetic fields, but often (particularly at high densities) these effects are insignificant.

The first equation of hydrodynamics is that of continuity, or the conservation of mass. Stated verbally, the rate at which mass accumulates in an element of volume is equal to the net rate at which it flows in through that element's boundaries. The rate of mass flow through any unit element of area is $\rho\vec{u} \cdot \hat{n}$, where ρ is the density, $\vec{u}$ is the velocity, and $\hat{n}$ is the unit normal to the surface. Using vector calculus, the net mass flow from an infinitesimal element of unit volume is given by $\nabla \cdot (\rho\vec{u})$, and the contained mass is ρ, so that the equation of continuity is

$$\frac{\partial \rho}{\partial t} + \nabla \cdot (\rho\vec{u}) = 0. \tag{3.1.1}$$

Although here ρ is the mass density, this equation is quite general; it applies to any conserved scalar quantity. Further, if a quantity is conserved except for known sources or sinks, a similar equation applies if these are added to the right hand side. Sources need not be anything as exotic as matter creation in a steady state cosmology, but can include gas "created" by the evaporation of a solid, produced in a chemical reaction, or introduced through a narrow pipe. Finally, if an equation like (3.1.1) applies to the individual components of a vector or tensor quantity, another similar equation applies to that entire quantity.

It is useful to consider a total (or convective) derivative, which measures the change in a quantity per unit time seen by an observer borne along with the fluid flow. This is defined

$$\frac{\mathcal{D}}{\mathcal{D}t} \equiv \frac{\partial}{\partial t} + \vec{u} \cdot \nabla. \tag{3.1.2}$$

The continuity equation may be rewritten

$$\frac{\mathcal{D}\rho}{\mathcal{D}t} + \rho \nabla \cdot \vec{u} = 0. \tag{3.1.3}$$

If ρ is constant the fluid is called incompressible, and $\nabla \cdot \vec{u} = 0$. This condition leads to a drastic simplification of the equations of hydrodynamics, and to a great body of mathematical results, but to very few astrophysical ones. Changes in density are important in astrophysics.

The second equation of hydrodynamics is equivalent to $F = ma$, or the conservation of momentum. An element of fluid feels a force which is given by the difference between the pressures on its surfaces. For an infinitesimal element the resulting equation is

$$\rho \frac{\mathcal{D}\vec{u}}{\mathcal{D}t} = -\nabla P, \tag{3.1.4}$$

where P is the pressure. Note that the convective derivative which follows the motion of a fluid element is used—at a fixed point in space $\vec{u}$ might change because fluid with different velocity is swept through the point ("is advected by the flow," in the jargon of fluid mechanics), so the partial time derivative of $\vec{u}$ contains contributions in addition to those of ∇P. Equation (3.1.4) may be rewritten in terms of the partial derivatives, using (3.1.2):

$$\frac{\partial \vec{u}}{\partial t} + (\vec{u} \cdot \nabla)\vec{u} = -\frac{1}{\rho}\nabla P. \tag{3.1.5}$$

In this form it is called Euler's equation. Viscous stresses and gravity and other body forces may be readily added, if present—they are "source terms" for momentum, in analogy with the sources of mass one might add to the continuity equation. Equation (3.1.5) may be rewritten, after some manipulation and the use of (3.1.1), in the form

$$\frac{\partial}{\partial t}(\rho\vec{u}) + \nabla\cdot(\rho\vec{u}\vec{u} + \mathbf{P}) = 0, \tag{3.1.6}$$

where $\vec{u}\vec{u}$ is a dyad and $\mathbf{P}$ is the stress tensor. For inviscid fluids $\mathbf{P}$ is the scalar pressure P multiplied by the unit tensor. Equation (3.1.6) is written in a form analogous to (3.1.1), so that it is obviously a conservation law for the momentum density $\rho\vec{u}$; $(\rho\vec{u}\vec{u} + \mathbf{P})$ is the momentum flux density tensor.

There generally is no independent hydrodynamic equation derivable from the conservation of angular momentum. Taking the cross product of a radius vector with the momentum equation would give an equation for the conservation of angular momentum, but this would contain no new information. An exception to this would occur for a fluid having internal stores of angular momentum other than its bulk motion (for example, particle spin), in which case there would be an additional equation relating the internal angular momentum to the angular momentum of the bulk motion.

The third equation of hydrodynamics is obtained from the conservation of energy. The first law of thermodynamics for a fluid element is

$$\frac{\mathcal{D}\mathcal{U}}{\mathcal{D}t} + P\frac{\mathcal{D}V}{\mathcal{D}t} = Q, \tag{3.1.7}$$

where $\mathcal{U} = \mathcal{E}/\rho$ is the internal energy per gram, $V \equiv 1/\rho$ is the specific volume, and Q is the external power supplied per gram (the source term); we now assume a scalar pressure P. Q includes such effects as thermal conduction, viscous frictional heating, radiation emission and absorption, and heat produced by chemical or nuclear reactions. Equation (3.1.7) may be rewritten, after some manipulation and the use of the continuity and momentum equations, in the form

$$\frac{\partial}{\partial t}\left[\mathcal{E}+\frac{1}{2}\rho u^2\right]+\nabla\cdot\left[\vec{u}\left(\mathcal{E}+\frac{1}{2}\rho u^2\right)+P\vec{u}\right]=\rho Q, \qquad (3.1.8)$$

paralleling equations (3.1.1) and (3.1.6) in explicit conservation form. It is apparent that $\mathcal{E}+\rho u^2/2$ is the energy density, and $P\vec{u}$ is the flux of mechanical work. The momentum and energy equations are formally identical to (3.1.1), with suitable definitions of the fluxes. This is as it should be, for they are all based on the conservation of quantities carried with the moving fluid.

These equations may be rewritten in a coordinate system which moves with the fluid, so that the independent space variables are replaced by masses. These coordinates, called Lagrangian (in contrast to the Eulerian coordinates of equations (3.1.1,5,6,8), are very useful in numerical computation, because the advective terms $\vec{u}\cdot\nabla$ are difficult to compute accurately. Lagrangian coordinates are particularly powerful in "one-dimensional" calculations, in which all material quantities depend on a single spatial coordinate. In these calculations, fluid cells or zones are generally infinitely broad flat slabs, infinitely long cylinders, or spherical shells. Two- and three-dimensional Lagrangian meshes (in which quantities depend on two or three spatial coordinates) are also used, but may tangle when a fluid flow is heavily sheared.

In Lagrangian coordinates the time derivative is $\mathcal{D}/\mathcal{D}t$, so the $\vec{u}\cdot\nabla$ terms do not appear explicitly and do not have to be computed. In one dimensional slab symmetry the spatial variable is the mass, with $dm\equiv\rho dx$. The equations become

$$\frac{\mathcal{D}V}{\mathcal{D}t}=\frac{\partial u}{\partial m}, \qquad (3.1.9)$$

$$\frac{\mathcal{D}u}{\mathcal{D}t}=-\frac{\partial P}{\partial m}, \qquad (3.1.10)$$

$$\frac{\mathcal{D}\mathcal{U}}{\mathcal{D}t}+P\frac{\mathcal{D}V}{\mathcal{D}t}=Q. \qquad (3.1.11)$$

The variables u, V, and $\mathcal{U}$ give the velocity, specific volume, and internal energy (per gram) of a specified fluid or mass element as a function of time. Eulerian coordinates may be obtained from

$$x(m,t) = \int_0^t u(m,t')\,dt' + x(m,0) = \int_0^m V(m',t)\,dm' + x(0,t). \tag{3.1.12}$$

Similar Lagrangian equations may be obtained for cylindrical and spherical geometry. Spherical geometry is of most astrophysical interest. The equations are:

$$dm = 4\pi r^2 \rho\, dr \tag{3.1.13}$$

$$\frac{\mathcal{D}V}{\mathcal{D}t} = \frac{\partial(4\pi r^2 u)}{\partial m} \tag{3.1.14}$$

$$\frac{\mathcal{D}u}{\mathcal{D}t} = -4\pi r^2 \frac{\partial P}{\partial m} \tag{3.1.15}$$

$$\frac{\mathcal{D}\mathcal{U}}{\mathcal{D}t} + P\frac{\mathcal{D}V}{\mathcal{D}t} = Q. \tag{3.1.16}$$

The radius r is obtained from

$$r(m,t) = \int_0^t u(m,t')\,dt' + r(m,0) = \int_0^m \frac{V(m',t)}{4\pi r^2}\,dm' + r(0,t). \tag{3.1.17}$$

A striking feature of the equations of hydrodynamics is their nonlinearity. This is apparent even in (3.1.1) in the product of $\vec{u}$ and ρ, but is true even for incompressible fluids because of the $(\vec{u}\cdot\nabla)\vec{u}$ term in (3.1.5). The practical consequence of this is that exact solutions are scarce. The most commonly used tools of the theoretical hydrodynamicist's trade are linearization of the equations for infinitesimal disturbances about a static equilibrium, and numerical computation of the full nonlinear equations.

The alert reader will have noticed that the first equation (3.1.1) involves two variables, ρ and $\vec{u}$. The number of scalar variables exceeds the number of equations by the number of spatial dimensions.

Adding the momentum equation (3.1.4) or (3.1.5) adds another variable P. This is a vector equation, with as many components as spatial dimensions, so now there is one more scalar variable than equations. Adding the energy equation (3.1.7) adds still another variable, $\mathcal{U}$ or $\mathcal{E}$ (assuming the source term is known). The excess of one variable remains. This is reminiscent of the closure problem encountered in radiation transport theory (**1.7**) and in kinetic theory (**2.1**). Its origin is similar, because in kinetic theory the hydrodynamic equations are obtained by taking velocity moments of the Boltzmann equation (2.1.9).

In order to close the system of equations an additional constraint is required. This is generally a constitutive relation, or "equation of state"

$$P = P(\mathcal{E}, V). \tag{3.1.18}$$

The justification for the use of (3.1.18) is the assumption of thermodynamic equilibrium. It may be possible to define the variables P and $\mathcal{E}$ in disequilibrium fluids, but only in equilibrium is there a unique relation among them. This relation is closely analogous to the Eddington approximation (1.7.12) of radiative transport theory, which has the form of an equation of state (3.1.18) for the radiation field. In equilibrium any two of the variables P, $\mathcal{E}$, and V are a complete description of the state of the fluid in its rest frame. Permeable or dielectric fluids in external fields, and intrinsically anisotropic fluids, require generalizations which are, in principle, straightforward; such complications are generally irrelevant to the astrophysicist.

To make thermodynamic equilibrium a valid approximation, it is necessary that the processes which bring it about occur either very rapidly, or very slowly, compared to the hydrodynamic processes of interest. A good example is a 100 Hz sound wave in air. Collisional relaxation between the molecules occurs in $\sim 10^{-9}$ seconds, and relaxation of the populations of most of the important rotational states is nearly that rapid, so the instantaneous achievement of equilibrium thermodynamic properties may be assumed. On the other hand, the nuclei are far from equilibrium; their lowest energy state is as ^{56}Fe nuclei. But nuclear reactions at room temperature are very slow ($\gg 10^{100}$ years), so the fact that ^{14}N and ^{16}O nuclei have other, more energetically favorable, states available may be ignored. Then thermodynamic equilibrium is well justified. But if the air contains a molecule whose rotational or vibrational relaxation time is 10^{-3} sec, then equilibrium is not obtained, and more complex equations than those of hydrodynamics must be solved. Such intermediate time scale processes do occur, and produce an observable excess attentuation of sound in air.

The equations of hydrodynamics have their ultimate derivation from the more complex "kinetic" equations, which deal with the complete distribution functions of the particle states. In attempting to solve these kinetic equations one generally takes their velocity moments, multiplying and integrating them $\int d^3\vec{u}\ (\vec{u}\vec{u}\cdots\vec{u})$, where the n-th moment equation is obtained if there are n factors in the parentheses. Each moment gives a new equation, but also an additional unknown moment of the distribution function. There is always an excess unknown, as we saw with the equations of hydrodynamics. In order to close this hierarchy of equations some additional assumption must be made. The section of this book **1.7** on radiation transport presents another example. In hydrodynamics the added assumption (3.1.18) is that of thermodynamic equilibrium. It is essential, and the term *hydrodynamics* is often taken implicitly to mean thermodynamic equilibrium. When similar equations arise there sometimes is no good way to close the hierarchy, as in the theory of turbulence, and uncertain approximations and assumptions must be made.

3.2 Sound Waves and Jeans Instability

Sound waves are one of the most important applications of the equations of hydrodynamics. In order to derive their properties, consider a one-dimensional disturbance of small amplitude in an infinite fluid which is otherwise uniform and at rest. Then we may write

$$\rho = \rho_0 + \delta\rho(x,t) \tag{3.2.1}$$

$$P = P_0 + \delta P(x,t) \tag{3.2.2}$$

$$u = \delta u(x,t). \tag{3.2.3}$$

Substitute these expressions into eqs. (3.1.1) and (3.1.5), and keep only terms of the first power in small quantities, to obtain a system of linear equations for small disturbances:

$$\frac{\partial \delta\rho}{\partial t} + \rho_0 \frac{\partial \delta u}{\partial x} = 0 \tag{3.2.4}$$

$$\frac{\partial \delta u}{\partial t} + \frac{1}{\rho_0}\frac{\partial \delta P}{\partial x} = 0. \tag{3.2.5}$$

To eliminate one of the three variables, write

$$\delta P = \left(\frac{\partial P}{\partial \rho}\right) \delta\rho. \tag{3.2.6}$$

This relation plays a role analogous to that of the constitutive relation (3.1.18) in closing the hierarchy of equations. Equation (3.2.6) may be regarded as a constitutive relation for the changes in pressure and density a fluid element undergoes in the sound wave, and may be derived from (3.1.18) if an additional assumption is made—for example, that processes in the sound wave are adiabatic.

In order to eliminate δu from (3.2.4) and (3.2.5) take the partial derivative with respect to time of (3.2.4) and that with respect to space of (3.2.5), and algebraically eliminate the cross-derivative term. The result is

$$\frac{\partial^2 \delta\rho}{\partial t^2} = \left(\frac{\partial P}{\partial \rho}\right) \frac{\partial^2 \delta\rho}{\partial x^2}. \tag{3.2.7}$$

If we define

$$c_s \equiv \left(\frac{\partial P}{\partial \rho}\right)^{1/2}, \tag{3.2.8}$$

the solutions to (3.2.7) are of the form

$$\delta\rho = f(x - c_s t) + g(x + c_s t), \tag{3.2.9}$$

where f and g are arbitrary functions. For a periodic function

$$\delta\rho = \delta\rho_0 \exp(ikx - i\omega t), \tag{3.2.10}$$

eq. (3.2.7) leads to

$$\omega^2 = c_s^2 k^2. \tag{3.2.11}$$

Thus c_s is the phase velocity (and also the group velocity $\frac{\partial \omega}{\partial k}$) of sound waves. (3.2.11) is called the dispersion relation (even though the waves are not dispersive).

Most commonly, the derivative in (3.2.8) is evaluated at constant entropy, because the fluid motions in sound waves are usually nearly adiabatic. Because we are concerned with infinitesimal disturbances, it is evaluated at the conditions of the undisturbed fluid. Then c_s is called the adiabatic sound speed.

Sound waves are not always adiabatic. For example, for short wavelength waves in a fluid of high thermal conductivity it may be a better approximation to evaluate (3.2.8) at constant temperature. Then c_s is an isothermal sound speed. For conditions intermediate between adiabatic and isothermal, sound waves are both dispersive and damped.

If the fluid satisfies the power law equation of state $P \propto \rho^\gamma$ for adiabatic changes, as is frequently a good approximation (for perfect monatomic gases $\gamma = 5/3$ for nonrelativistic point particles and $\gamma = 4/3$ if the particles are relativistic; see **1.9**), then $c_s = \sqrt{\gamma P/\rho}$. If, in addition, $P = \rho N_A k_B T/\mu$, where μ is the molecular weight (as is the case for gases of noninteracting particles), then

$$c_s = \sqrt{\frac{\gamma N_A k_B T}{\mu}}. \tag{3.2.12}$$

The most important qualitative implication of (3.2.12) is the increase of c_s with T.

For the diatomic molecules in air at room temperature the vibrational degrees of freedom are not significantly excited (the first excited state has an energy corresponding to 3340 °K in N_2 and 2230 °K in O_2), while the rotational degrees of freedom are excited to such high quantum numbers that they may be considered classical oscillators (the first excited states are at energies corresponding to 6 °K and 4 °K respectively). Then there are effectively 5 degrees of freedom per molecule, and (3.2.12) with $\gamma = 1.40$ (1.9.8) is an excellent approximation to the sound speed in air.

We now consider sound waves in a self-gravitating fluid. To the momentum equation we must add the force of gravity, which we express in terms of the gravitational potential ϕ:

$$\frac{\partial \vec{u}}{\partial t} + (\vec{u} \cdot \nabla)\vec{u} = -\frac{1}{\rho}\nabla P - \nabla\phi. \tag{3.2.13}$$

The potential is given by Poisson's equation

$$\nabla^2\phi = 4\pi G\rho. \tag{3.2.14}$$

As before, we consider small perturbations about a uniform infinite fluid at rest.

This procedure is incorrect for a self-gravitating fluid, because the assumed initial uniform state is not a solution of the equations;

(3.2.14) cannot be solved for such an infinite fluid without having $\phi \to \infty$ and $\nabla\phi \to \infty$, implying an infinite gravitational acceleration! In fact, (3.2.13) cannot have a uniform static equilibrium solution even for a finite self-gravitating fluid, because a uniform fluid possesses no pressure gradient to oppose the acceleration of gravity. Although technically improper, the calculation is still informative.

For small disturbances $\delta\rho(x,t)$, $\delta P(x,t)$, $\delta u(x,t)$, and $\delta\phi(x,t)$, (3.2.13) and (3.2.14) become

$$\frac{\partial \delta\rho}{\partial t} + \frac{1}{\rho_0}\frac{\partial \delta P}{\partial x} + \frac{\partial \delta\phi}{\partial x} = 0 \tag{3.2.15}$$

$$\frac{\partial^2 \delta\phi}{\partial x^2} - 4\pi G\delta\rho = 0 \tag{3.2.16}$$

(3.2.15) replaces (3.2.5). As before, we assume that all the small quantities have spatial and temporal variability $\propto \exp(ikx - i\omega t)$, and eliminate $\delta\rho$, δP, and $\delta\phi$ from equations (3.2.4), (3.2.6), (3.2.15) and (3.2.16). The result (if all the small quantities are nonzero) is an algebraic dispersion relation:

$$\omega^2 = c_s^2 k^2 - 4\pi G\rho_0. \tag{3.2.17}$$

The important property of (3.2.17) is the existence of a critical wavevector k_J and wavelength λ_J:

$$k_J = \sqrt{4\pi G\rho_0}/c_s \tag{3.2.18a}$$

$$\lambda_J = 2\pi c_s/\sqrt{4\pi G\rho_0} \tag{3.2.18b}$$

For $k \gg k_J$ the second term on the right hand side of (3.2.17) is negligible, and it closely approximates (3.2.11); these disturbances are essentially ordinary sound waves, and self-gravity is unimportant. For $k < k_J$ we have $\omega^2 < 0$, so there are both exponentially growing (in time) and exponentially damped solutions. The growing solutions describe Jeans instability. Their growth rate is $\sim \sqrt{4\pi G\rho_0}$.

If this derivation had begun from consistent initial conditions, the Jeans instability criterion would set an upper bound λ_J on the size of a stable self-gravitating gas cloud. Because it did not, it is better to regard λ_J as a parameter dividing length scales for which self-gravity is important from those for which it is not. A cloud of radius $r < \lambda_J$ will, if unconfined, expand under the influence of its own pressure forces. If it is to survive for a time $\gtrsim r/c_s$ it must be confined by an external pressure; its self-gravity is inadequate. This is believed to be a qualitatively correct description of most interstellar gas clouds, which have $r \ll \lambda_J$, and which are believed to be immersed in a dilute but hot medium with which they are in pressure balance.

In a cloud of radius $r > \lambda_J$ the force of gravity exceeds that of pressure, and the Jeans analysis predicts gravitational collapse in a time $\sim (4\pi G\rho_0)^{-1/2}$. If such a collapse is followed into the nonlinear regime, the pressure forces are found (if $\gamma > 4/3$) to increase more rapidly that the gravitational forces, and the cloud settles into (or oscillates around) a state of hydrostatic equilibrium. In this state, which resembles the hydrostatic equilibrium of a star, $r \sim \lambda_J$ provided the new, increased, ρ and c_s are used in the calculation of λ_J. The frequencies of the low modes of oscillation are $\sim \sqrt{4\pi G\rho} \sim t_h^{-1}$, where t_h is the hydrodynamic time scale (1.6.1).

Because of the improper assumptions made in the derivation of λ_J it is best to regard it only as a criterion for the importance of self-gravity. In objects with $r < \lambda_J$ self-gravity is negligible, while in those with $r \sim \lambda_J$ it is important. For example, an interstellar cloud which grows to this size will then collapse.

The analysis is more complicated if the fluid has internal motions in its initial state. Then it is qualitatively correct to replace c_s by a characteristic fluid velocity v. We may smoothly interpolate between the $c_s \gg v$ and $c_s \ll v$ limits with an expression like

$$\lambda_J = 2\pi \left(\frac{c_s^2 + v^2}{4\pi G \rho_0} \right)^{1/2} ; \tag{3.2.19}$$

the particular form of the numerator is suggested by the addition of the hydrodynamic stress ρv^2 to the gas pressure, but is not quantitatively justified. More importantly, the appropriate value of v will almost always depend on the length scale being considered, because it is only the *differential* velocity across a distance r which can prevent collapse (it is always possible to consider the problem in the center-of-mass frame of the region under consideration).

For a turbulent flow the differential $v(r)$ depends on the statistical properties of the turbulence, and will generally be an increasing function of r. If the turbulent velocities are subsonic, then v makes a minor contribution to (3.2.19), whatever its dependence on r. In supersonic turbulence fluid elements collide and strong shocks form (**3.3**), increasing the sound speed so that $c_s \sim v$.

An important case of differential motion is that of rotation, such as is found in a disc of gas orbiting a central object. Examples include the interstellar gas orbiting the centers of galaxies, and accretion discs surrounding stars and collapsed objects. In these cases $v = \Omega r$, where Ω is the local angular velocity of rotation. Substituting this into (3.2.19), neglecting c_s, and using $\Omega = \sqrt{GM/R^3}$, where M and R are the radius and mass of the central object (this holds approximately even if the mass is distributed in a non-spherically symmetric way throughout a volume of radius R, as is the mass of galaxies), we find that $r > \lambda_J$ requires

$$\rho_0 > \pi \frac{M}{R^3}. \tag{3.2.20}$$

The numerical coefficient is rather approximate (because it depends on the boundary conditions on the disturbances, which we have ignored). The qualitative conclusion is that discs much denser than the mean density of their enclosed mass may be unstable. Because $c_s > 0$ it is also necessary that $r > \lambda_J$ where λ_J is obtained from (3.2.18b); to be unstable a rotating cloud must have both a minimum size and a minimum density.

For the mean interstellar medium near us $\rho_0 \sim 10^{-24}$ gm/cm^3, while $M/R^3 \sim 10^{-23}$ gm/cm^3, so that instability is not expected. In much denser clouds of sufficient size instability will occur; this is probably how stars and star clusters begin to form. During their subsequent collapse angular momentum is conserved, so that $v \propto r^{-1}$ and $\rho_0 \propto r^{-3}$; r decreases faster than $\lambda_J \propto r^{1/2}$ (3.2.19), halting the collapse. Further contraction is limited by the rate at which other processes (probably magnetic torques) can remove angular momentum from this protocluster or protostar. These processes are incalculable, and may be very slow.

Accretion discs are usually (but uncertainly) estimated to have very little mass and low density, and therefore not to satisfy the instability criterion (3.2.20).

A closely related problem is the Jeans instability of a gas of collisionless gravitating particles, such as the stars in a galaxy, or the denser clouds (which may be protostars) formed by the Jeans instability in a more dilute gaseous medium. The analysis is more complicated because the equations of hydrodynamics cannot be used to describe the motions of collisionless particles, but the instability criterion is similar, if c_s is now taken to be the velocity dispersion of the particles. The stars in the disc of our Galaxy have $\rho_0 \sim 10^{-23}$ gm/cm^3, and are thus closer to instability than the gas alone. Because the particles of a collisionless gas pass by each other freely, without exchanging momentum or "sticking," one consequence of Jeans instability is an increase in their velocity dispersion as δu is added to the pre-existing velocities. It may be that this process maintains the velocity dispersion, and limits the density, of the stars in a galactic disc. It is also possible that spiral arms are a consequence of density clumping produced by Jeans instability of the disc, stretched into a spiral pattern by galactic differential rotation.

In a roughly spherical object, whether collisional (a star), or collisionless (an elliptical galaxy or star cluster), $\rho_0 \sim M/R^3$ and c_s (or v) $\sim \sqrt{GM/R}$, so that $R < \lambda_J$, and there is no instability.

3.3 Shocks

A shock is a propagating irreversible discontinuity in the thermodynamic state of a body of material. For astrophysical purposes we are interested only in shocks in fluids, though they can occur in solids. We will assume that everywhere in the fluid, except in the infinitesimally thin sheet which we call the shock, thermodynamic equilibrium holds, and the fluid is locally characterized by its velocity, density, pressure, and internal energy. As discussed in the previous section, in thermodynamic equilibrium these are a complete description of the fluid, and an additional constitutive relation makes one of the variables redundant.

We are interested in finding relations among the parameters of the unshocked fluid ("ahead" of the shock), those of the shocked fluid ("behind" the shock), and the velocity of propagation of the shock itself. These relations, often referred to as the "jump conditions," are determined by the conservation laws.

It is not *a priori* obvious how a shock may actually be produced. It is even less obvious (and quite a difficult problem to determine) how unshocked fluid is transformed to its shocked state. The relations among the parameters of the shocked and unshocked fluid are only a restatement of the conservation laws, and do not answer these questions.

Imagine a fluid initially at rest with uniform density ρ_0 and pressure P_0. At $t = 0$ we consider a shock propagating in the $+x$

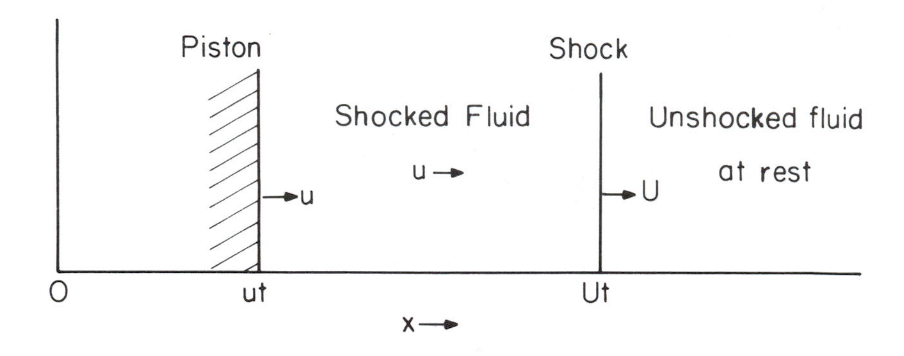

Figure 3.1. Shock produced by a piston.

direction from the origin with a velocity U. The gas behind the shock moves with a velocity u, and has a uniform density ρ_1 and pressure P_1 throughout its volume. The equations of hydrodynamics are satisfied everywhere but at the shock front, for everywhere else $\partial/\partial t = \partial/\partial x = 0$. This configuration (Figure 3.1) may be produced by a piston beginning to move at $t = 0$ from the origin with speed u, so that the shocked gas is at rest with respect to the piston.

Conservation of mass gives the equation

$$\rho_1(U - u)t = \rho_0 U t, \tag{3.3.1}$$

where the right hand side is the rate of mass flow into the shock front, and the left hand side is the rate of mass flow out of it. No mass can accumulate in the infinitesimally thin shock.

Conservation of momentum gives

$$\rho_0 Utu = (P_1 - P_0)t, \qquad (3.3.2)$$

where $P_1 - P_0$ is the pressure jump across the shock. Without specifying the internal workings of the shock, we can say that matter can acquire momentum only by the application of a pressure gradient or pressure drop. Alternatively, we may note that the piston at the left must apply a pressure P_1, balancing the fluid pressure upon its face, and a hypothetical piston at $x = +\infty$ must apply a pressure P_0. These imply a change in momentum per unit area per unit time $P_1 - P_0$, which appears in the momentum matter acquires by changing its velocity as it crosses the shock.

Conservation of energy gives

$$\rho_0 Ut \left(\mathcal{U}_1 - \mathcal{U}_0 + \frac{1}{2}u^2 \right) = P_1 ut, \qquad (3.3.3)$$

which is the rate at which the piston at the left does work on the fluid, measured in the laboratory frame. In order to keep the shock propagating at a constant rate, it is necessary to supply continually energy and momentum from the outside. If this is not done then our simple solution, uniform everywhere but at the shock, will be replaced by a more complex solution in which the shock gradually weakens.

Now make a Galilean transformation to a frame moving to the right at a speed U, so that the shock is stationary. In this frame the unshocked matter has a velocity

$$u_0 = -U, \qquad (3.3.4)$$

and the shocked matter has a velocity

$$u_1 = u - U. \qquad (3.3.5)$$

Equations (3.3.1)–(3.3.3) may then be rewritten, after a little algebra:

$$\rho_1 u_1 = \rho_0 u_0 \tag{3.3.6}$$

$$P_1 + \rho_1 u_1^2 = P_0 + \rho_0 u_0^2 \tag{3.3.7}$$

$$\mathcal{U}_1 + \frac{P_1}{\rho_1} + \frac{1}{2} u_1^2 = \mathcal{U}_0 + \frac{P_0}{\rho_0} + \frac{1}{2} u_0^2. \tag{3.3.8}$$

These equations bear a striking resemblance to the equations of hydrodynamics in conservation law form, (3.1.1), (3.1.6), and (3.1.8). This should be no surprise, because both sets of equations only state the basic conservation laws. To derive (3.3.6)–(3.3.8) from the hydrodynamic equations, integrate each of the latter over a small spatial region including the shock front, working in the frame in which the shock is stationary. As the thickness of this region becomes infinitesimal, $\int dx \frac{\partial}{\partial t}$ goes to zero, because $\partial/\partial t$ is finite everywhere. At the shock $\partial/\partial x$ is a δ-function, but $\partial/\partial t$, although not defined, is not a δ-function because no variable exhibits a discontinuous change in time. (For flows uniform on each side of the shock $\partial/\partial t = 0$; integrating over an infinitesimal thickness generalizes the conclusion to flows whose properties vary smoothly away from the shock.) Then $\int dx \frac{\partial f(x)}{\partial x}$ across the shock reduces to $f(x_1) - f(x_0)$, and if there are no δ-function sources at the shock the result is $f(x_1) = f(x_0)$ for the three functions f of (3.1.1), (3.1.6), and (3.1.8). This gives (3.3.6)–(3.3.8); to obtain the last of these divide the equation for the third f by that for the first one.

From (3.3.6) and (3.3.7) we readily obtain

$$u_0^2 = \frac{\rho_1}{\rho_0} \left(\frac{P_1 - P_0}{\rho_1 - \rho_0} \right). \tag{3.3.9}$$

This has a simple but interesting interpretation. We know that the speed of an infinitesimal sound wave in a fluid is $\sqrt{\partial P/\partial \rho}$. The ratio

$(P_1 - P_0)/(\rho_1 - \rho_0)$ is an approximation to $\partial P/\partial \rho$, and should lie between the values of $\partial P/\partial \rho$ for the unshocked and shocked fluid. The sound speed in the heated, compressed, shocked fluid is larger than in the unshocked fluid, so we may generally take $c_1^2 > (P_1 - P_0)/(\rho_1 - \rho_0) > c_0^2$, where c denotes the sound speed. We also have $\rho_1 > \rho_0$, so that (3.3.9) gives

$$u_0^2 > c_0^2. \tag{3.3.10}$$

The shock advances into the unshocked fluid at a speed in excess of its sound speed (supersonically); shocks have Mach numbers greater than 1.

Elementary algebra (or an interchange of indices in 3.3.9, for the original equations are symmetric in them) leads to

$$u_1^2 = \frac{\rho_0}{\rho_1}\left(\frac{P_1 - P_0}{\rho_1 - \rho_0}\right). \tag{3.3.11}$$

By essentially the same argument as before this leads to

$$u_1^2 < c_1^2. \tag{3.3.12}$$

The shocked matter moves away from the shock at a speed less than its sound speed (shocks are subsonic with respect to the matter behind them).

Equations (3.3.6)-(3.3.8) are three equations relating the four "unknowns" ρ_1, u_1, P_1, $\mathcal{U}_1$ to the "knowns" ρ_0, u_0, P_0, $\mathcal{U}_0$. If there is an additional constraint, it is possible to solve explicitly for the unknowns in terms of the knowns. Because we have assumed that the fluid is in thermodynamic equilibrium on each side of the shock, we are assured that such an additional constraint exists in the form of an equation of state, eq. (3.1.18).

Take the perfect gas equation of state

$$\mathcal{E} = \frac{1}{\gamma - 1} P, \tag{3.3.13}$$

where γ is a constant. For reversible adiabatic transformations we showed in **1.9** that such a gas follows the law

$$P \propto \rho^{\gamma}. \tag{3.3.14}$$

This is a convenient and often useful law, but we must remember that shocks are not reversible, and do not follow (3.3.14), even if the gas satisfies (3.3.13). In this section we use (3.3.13), and only mention (3.3.14) to remind the reader of the significance of γ.

The jump conditions give

$$\mathcal{U}_1 - \mathcal{U}_0 = \frac{1}{2}(P_1 + P_0)(V_0 - V_1), \tag{3.3.15}$$

which neatly expresses the change in $\mathcal{U}$ as a result of the mean "PdV" work. Using the equation of state (3.3.13) we obtain

$$\frac{P_1}{P_0} = \frac{(\gamma+1)V_0 - (\gamma-1)V_1}{(\gamma+1)V_1 - (\gamma-1)V_0} \tag{3.3.16a}$$

or

$$\frac{V_1}{V_0} = \frac{(\gamma-1)P_1 + (\gamma+1)P_0}{(\gamma+1)P_1 + (\gamma-1)P_0}. \tag{3.3.16b}$$

Equation (3.3.16) has a striking and important consequence. In the limit $P_1/P_0 \to \infty$ we have

$$\frac{V_1}{V_0} \to \frac{\gamma-1}{\gamma+1}. \tag{3.3.17}$$

No single shock, no matter how strong, can compress matter by a factor of more than $(\gamma+1)/(\gamma-1)$, which is equal to 4 for a simple monatomic gas with $\gamma = 5/3$. The pressure and temperature of the shocked matter may increase without bound, but the density cannot. The only exception to this rule is obtained if $\gamma = 1$. This describes an isothermal gas, for which the temperature is fixed, but whose density may be increased arbitrarily by a single shock.

A practical realization of an isothermal gas is one which cools radiatively, and for which the radiative cooling rate is a very steeply increasing function of temperature. This is, in fact, often a good approximation for the interstellar gas, because its cooling depends on collisional excitation of energy levels with excitation energy $\chi \gg k_B T$. The cooling rate then varies approximately as $\exp(-\chi/k_B T)$, with a very large coefficient. The consequence of this is that strong shocks in the interstellar medium (such as those produced by supernova explosions) produce sheets or shells of very high density, at least in calculations.

These dense shells may be sites of star formation. This suggestion is controversial, in part because the compression is in one dimension only. The Jeans length of a flattened cloud is approximately determined by its mean density averaged over a spherical volume λ_J^3; it is easily seen that the flattening to a pancake of a cloud initially smaller than λ_J will not, by itself, make it gravitationally unstable. The increased confining pressure behind the shock will contribute to instability. The interstellar medium has a very heterogeneous density distribution, so that different parts of the shell will travel at different speeds, depending on the density of the material they encounter. Transverse density gradients will also alter their direction of motion (refracting the shock). Fragments of the dense shell may acquire a significant velocity dispersion, which interferes with gravitational collapse (**3.2**).

Sonic booms are familiar examples of weak shocks. The pressure jump $P_1 - P_0$ is orders of magnitude less than P. Writing $P_1 = P_0(1+\alpha)$, and taking $\alpha \ll 1$ in (3.3.16) leads to $\rho_1 \approx \rho_0(1+\alpha/\gamma)$. Weak shocks therefore satisfy

$$\frac{d\ln P}{d\ln \rho} = \gamma, \tag{3.3.18}$$

the same equation that applies to infinitesimal sound waves in a gas with equation of state (3.3.13). By (3.3.9) and (3.3.11), a weak shock moves at the sound speed with respect to both the shocked and unshocked gas, and in all respects resembles a sound wave of impulsive form. The frequency spectrum is given by the Fourier transform of its step function pressure profile.

Shocks may be produced in a wide variety of circumstances. The piston discussed at the beginning of this section is only the simplest possibility. If a fluid contains a propagating pressure profile resembling Figure 3.2.a, the sound speed will be higher in the adiabatically compressed and heated matter near the crest of the wave than elsewhere. For this reason, and because the nonlinear $(\vec{u} \cdot \nabla)\vec{u}$ term has a similar effect, the peak of the pressure profile will gradually overtake its leading edge, and its shape will steepen, as shown in Figure 3.2.b. Eventually a shock will form, as shown in Figure 3.2.c. This qualitative description is confirmed by numerical calculations.

Sound waves of infinitesimal amplitude propagate at constant speed with a stationary profile. The amplitude of any real disturbance is not infinitesimal, so it might be expected to steepen and form a shock, as sketched in Figure 3.2. This usually does not happen. The steepening is slow if the amplitude is small. For example, the sounds of ordinary speech have an intensity of about 75 dB in the jargon of the acoustic engineer, who defines the intensity of sound as $20 \log_{10}(\delta P/2 \times 10^{-10}$ atmospheres) dB. They would require propagation through a distance $D \sim 10^6$ wavelengths to steepen substantially, even if there were no spherical divergence. Unless a sound wave is initially quite strong, its steepening is usually overwhelmed by losses resulting from viscosity, heat conduction, and delayed relaxation to thermodynamic equilibrium (bulk viscosity). Sound waves are also significantly attenuated at boundaries between fluid media and solids, or between two fluids, because there are usually dissipative heat flows or viscous forces at interfaces.

Sound waves share with shallow water waves the property that the velocity of propagation increases at the crest of the wave. As a re-

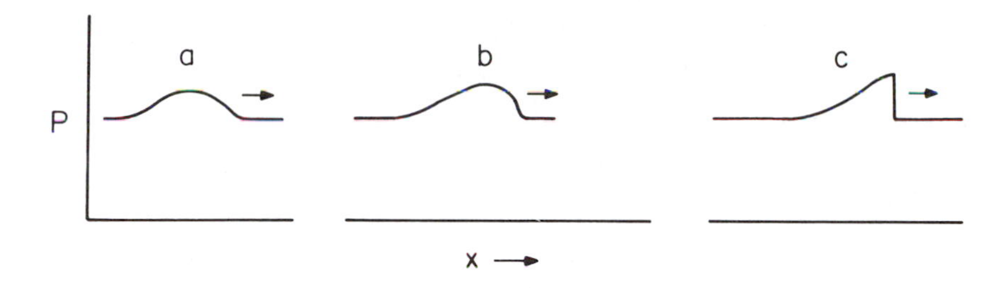

Figure 3.2. Shock formation.

sult, shallow water waves also show a steepening process qualitatively resembling that of Figure 3.2. The "shock" is called a hydrodynamic bore, and is observed on beaches as water runs up them after a wave breaks, in some tidal estuaries, and in other circumstances where a large surge of water is suddenly released. The resemblance is only qualitative, for the dispersion relation for infinitesimal shallow water waves is dispersive. As a result, there exist stationary nondissipative propagating structures of finite amplitude, called solitons, in which the dispersion is balanced by the nonlinearity. Acoustic solitons are not possible because there is no dispersion to balance their tendency to steepen into a shock.

We have said nothing about what actually goes on in a shock;

that is, about the nonequilibrium processes by which the material is transformed from its equilibrium unshocked state to its equilibrium shocked state. Weak shocks may, in some circumstances, be calculated as if the matter were always close to equilibrium, so that ordinary coefficients of viscosity and thermal conductivity (whose derivation assumes small deviations from equilibrium) may be used. A strong shock is a region of large velocity and thermal gradients, about a mean free path thick, in which deviations from thermodynamic equilibrium are large, and the material undergoes an irreversible change of state. The hydrodynamic description of a fluid is not valid there, and the transport coefficients (viscosity and thermal conductivity) are not well defined. To compute the structure of a strong shock quantitatively requires consideration of the full nonequilibrium particle distribution function, and is quite difficult. Various additional complications are possible, including "collisionless" plasma shocks, in which dissipation is provided by plasma instabilities (in effect, macro-collisions) rather than by microscopic particle collisions. Fortunately, for most purposes it is possible to ignore the detailed structure of the shock itself. The power of the assumption of thermodynamic equilibrium outside of the shock is that it permits calculation of all the properties of the fluid from the conservation laws alone, without any consideration of the way in which the equilibrium states are achieved.

3.4 Blast Waves

An explosion or other instantaneous point release of energy within a fluid produces an outward travelling shock. This is called a blast wave. A full description of the flow is very complex, but if certain assumptions can be justified a simple theory is both accurate and informative. The theory was first developed in order to describe man-made explosions in air, and is known as the Sedov-Taylor solution, but also has astrophysical applications, particularly to supernova explosions in the interstellar medium. Zeldovich and Raizer (1966) give a thorough discussion with references to the original literature.

We first assume that the fluid medium is uniform and at rest. This is well justified for most atmospheric explosions (so long as the radius of the shock is much less than the scale height of the atmosphere, and the shock has not touched the ground), but is more questionable for interstellar explosions. The interstellar medium is strongly heterogeneous, with its density believed to vary by a factor of at least $\sim 10^3$ between the abundant ordinary clouds and the hot intercloud medium (rarer dense clouds may be orders of magnitude denser still). It is thus necessary to assume either that an average may be taken over the heterogeneities (combining high and low density regions to form an effective mean medium), or that the presence of the clouds may be ignored (if they do not affect the flow around them, or if they are sparsely enough distributed that a typical blast wave does not encounter them). The validity of any of these assumptions of the interstellar medium has not been established. Many interstellar blast waves produced by supernovae (called supernova remnants) are observed, but tests of the theory are indirect and uncertain because its basic parameters—the age and energy of the explosion and the density of the medium—are generally not directly observable or quantitatively known.

This problem is described by several parameters: the energy Y and mass M of the exploding matter, and the density ρ_0 and pressure P_0 of the medium. We always assume $Y/M \gg P_0/\rho_0$; if this were not the case the explosion would never produce a strong shock. Three regimes may be distinguished on the basis of the radius R of the spherical blast wave:

In the early regime I, $R \ll (M/\rho_0)^{1/3}$, and the mass contained within the radius R is almost entirely that of the exploding object. The surrounding medium has had little influence on its motion. Supernova remnants typically remain in this regime for the first $\sim 300 - 3000$ years of their lives, depending on M, ρ_0, and their initial velocity of explosion. The Crab Nebula supernova remnant, the relic of a supernova of the year 1054, is believed to be in this regime, in large part because for it ρ_0 and the initial expansion velocity were unusually low.

In the late regime III, $R \gg (Y/P_0)^{1/3}$. Most of the energy contained within the blast wave was internal energy of the fluid prior to the explosion, with Y contributing only a small fraction. The shock is now weak and propagates at nearly the sound speed.

The middle regime II is the most interesting one; here $(Y/P_0)^{1/3} \gg R \gg (M/\rho_0)^{1/3}$. In this case most of the mass within the blast wave was that of the fluid medium, but most of the energy was that of the explosion. If these inequalities hold to high accuracy then P_0 and M may be ignored, and the problem is described by only two parameters, Y and ρ_0, which permits great simplification. We first discuss the early part IIa of this middle regime, in which radiative losses are negligible, the total energy within the blast wave is essentially constant (and equal to Y), and the fluid is described by an adiabatic exponent $\gamma > 1$.

All physical quantities are products of powers of mass, length, and time. If a problem has three intrinsic dimensional parameters of independent dimensionality (meaning that none is proportional to a product of powers of the other two; equivalently, the matrix formed by the powers of mass, length, and time in the three quantities is not singular), then it is possible to define a characteristic length (or any other dimensional quantity) as a product of appropriate powers of the dimensional parameters. The estimates of **1.4** were examples of this procedure; a characteristic density ρ, pressure P, and thermal energy $k_B T^*$ were found for a star as a function of its parameters G, M, and R.

If only two independent dimensional parameters exist then, in general, no characteristic length or time may be defined (some characteristic dimensional parameters may be formed from the two given parameters, but in our case not the important ones of length, time, or velocity). The solutions are of the form

$$\begin{aligned} P(r,t) &= P_1(t)\tilde{P}(r/R(t)) \\ u(r,t) &= u_1(t)\tilde{u}(r/R(t)) \\ \rho(r,t) &= \rho_1(t)\tilde{\rho}(r/R(t)), \end{aligned} \tag{3.4.1}$$

where the quantities P_1, u_1, and ρ_1 are the pressure, velocity, and density immediately behind the shock, $\tilde{P}$, $\tilde{u}$, and $\tilde{\rho}$ are dimensionless functions of dimensionless arguments, and $R(t)$ gives the blast wave radius as a function of time. The form (3.4.1) is called a similarity solution because the solution at any one time looks like that at any other time if all lengths are multiplied by a single scale factor; both solutions have the same shape. Such a solution is possible only if the intrinsic parameters of the problem define no characteristic lengths or times, for if they did the solutions when R or t were near these characteristic lengths or times could be of a different form than those when R or t were much greater or much less.

Our assumption that there are no dimensional parameters other than Y and ρ_0 requires that the constitutive relations be of the form (3.3.13) and (3.3.14), because any other form would contain additional dimensional parameters. Because P_0 is negligible the strong shock jump conditions (**3.3**) may be used to express $P_1(t)$, $u_1(t)$, and $\rho_1(t)$ in terms of the shock velocity; $\rho_1(t)$ is the constant density

---

* The use of the temperature T in place of $k_B T$ is only a redefinition of the temperature scale for the sake of convenience; °K is not a physical unit independent of energy.

$\rho_0(\gamma+1)/(\gamma-1) \geq 4\rho_0$. Because $\rho_1 \gg \rho_0$ most of the swept up mass is concentrated in a dense shell just inside the blast wave.

Equations (3.4.1) may be substituted in the hydrodynamic equations in spherical geometry in order to obtain ordinary differential equations for $\tilde{P}$, $\tilde{u}$, and $\tilde{\rho}$. It is more interesting and important to find the function $R(t)$, and a simple dimensional argument suffices. If we consider the blast wave at a specific time t, this provides a third dimensional parameter. From Y, ρ_0, and t we can construct one quantity with the dimensions of length. Because there are also no dimensionless parameters (pure numbers) in the problem other than γ, $R(t)$ must be proportional to this one characteristic length:

$$R(t) = \xi_0(\gamma)\left(\frac{Yt^2}{\rho_0}\right)^{1/5}. \tag{3.4.2}$$

The dimensionless function $\xi_0(\gamma)$ is the constant of proportionality, and is found by integration of the equations for $\tilde{P}$, $\tilde{u}$, and $\tilde{\rho}$; it is fairly close to unity.

From (3.4.2) we obtain the shock velocity

$$\frac{dR(t)}{dt} = \frac{2}{5}\xi_0(\gamma)\left(\frac{Y}{\rho_0 t^3}\right)^{1/5}. \tag{3.4.3}$$

The functions $P_1(t)$ and $u_1(t)$ may be obtained from (3.4.3) and the strong shock jump conditions; on dimensional grounds alone we have

$$u_1(t) \sim \frac{dR(t)}{dt} \propto \left(\frac{Y}{\rho_0}\right)^{1/5} t^{-3/5} \propto \left(\frac{Y}{\rho_0}\right)^{1/2} R^{-3/2} \tag{3.4.4a}$$

$$P_1(t) \sim \rho_0\left(\frac{dR(t)}{dt}\right)^2 \propto Y^{2/5}\rho_0^{3/5}t^{-6/5} \propto YR^{-3} \tag{3.4.4b}$$

Equations (3.4.2)–(3.4.4) are used in theories of supernova remnants. Because they are believed to be typically 10,000 – 30,000 years old, their ages and the time-dependence of their properties have not been directly observed, and comparisons with the theory are only statistical. The numerical values of Y and ρ_0 are also poorly known, and perhaps the most important application of this theory is to determine the ratio Y/ρ_0 from (3.4.2). Typical values obtained from statistical analyses are $\sim 3 \times 10^{75}$ cm^5/sec^2 (Clark and Caswell 1976). There are at least two distinct types of supernovae and the interstellar ρ_0 is known to be very heterogeneous, so it is not clear how a mean Y/ρ_0 should be interpreted.

In the early part of regime II, just described, the temperature of the shocked matter (which is proportional to $P_1/\rho_1 \propto t^{-6/5} \propto R^{-3}$) exceeds 10^6 °K, and radiative cooling is unimportant. However, the radiative cooling rate of interstellar matter (Spitzer 1978) rises steeply when T drops below 10^6 °K.

We now consider regime IIb, in which $(Y/P_0)^{1/3} \gg R$ still holds, but radiative energy losses are rapid, and the matter may be approximately described as an isothermal fluid (the adiabatic exponent $\gamma \to 1$). Now $\rho_1 \to \infty$, and the swept up mass forms a very thin and dense shell just behind the blast wave. A simple "snowplow" model is useful.

Because radiative cooling is effective, the pressure in the low density region inside the dense shell is very low. Each element of solid angle of the shell now moves independently of all other elements, and is slowed as it sweeps up further interstellar material. Its collision with the interstellar matter is completely inelastic because of the rapidity of the radiative losses; momentum is conserved but all the kinetic energy (in the center-of-mass frame of each element) is radiated. If we enter regime IIb at time t_0 with a shell of radius R_0, velocity u_0, and mass per unit area $\sigma_0 = R_0\rho_0/3$, its further slowing is described by

$$u = u_0 \left(\frac{R_0}{R}\right)^3 \qquad (3.4.5a)$$

$$R \doteq \left(R_0^4 + 4(t - t_0)u_0 R_0^3\right)^{1/4}. \tag{3.4.5b}$$

When u calculated from (3.4.5a) becomes comparable to c_s, the blast wave enters the weak shock regime III. This is discussed in Appendix B.

3.5 Accretion

Accretion is the name given the process by which an object increases its mass by the capture of surrounding matter. We are here concerned with the gravitational capture of gas by stars, and not with the agglomeration of solid particles (for example, in the early Solar System), in which surface forces are important. In this section we consider only the processes by which gravitationally unbound matter is captured, and not its subsequent flow onto the massive object which attracted it; these flows are described in **3.6** and **3.7**.

Accretion was first investigated by astronomers who were concerned with the possibility that the masses of stars might grow appreciably during their lifetimes by accretion of interstellar material, or that the potential energy released by infall onto the Sun might affect the climate of the Earth. The development of a quantitative theory of accretion and better estimates of the density of the interstellar medium demonstrated that accretion is almost always insufficient to produce these effects, although between a protostar forming from a dense interstellar cloud and a star is a state which may be described as a rapidly accreting star.

Modern interest in accretion focuses on two distinct problems. The first concerns stars (including neutron stars and black holes) in close binaries, which may accrete significant amounts of mass from their companions. In some cases this mass flows smoothly from one star to the other, and is always gravitationally bound, but in others it must be captured from a high velocity wind (like those discussed in **1.15**); in the latter case the dynamics of accretion resemble that of accretion from the interstellar medium. Accretion onto collapsed stars is of interest because a great deal of gravitational potential energy is released. The second problem concerns peculiar stars, particularly white dwarves, whose surface composition and spectrum may be observably affected by the accretion of very small amounts of interstellar material.

Consider accretion by a star of mass M and radius R moving at a velocity v through an initially uniform medium of density ρ_0. Assume

$$v \ll \sqrt{GM/R}, \tag{3.5.1}$$

as is usually the case, and first take the particles of the medium to be collisionless. Then each particle follows the hyperbolic path of a free test particle, unless it actually collides with the stellar surface. It is simple to calculate the limiting impact parameter b_1, within which particles collide with the star and are accreted. Particles with this impact parameter have orbits just tangent to the stellar surface. Because of (3.5.1) their orbits are nearly parabolic, and by conservation of angular momentum we have

$$b_1 v = \sqrt{2GMR}. \tag{3.5.2}$$

The accretion rate A is then given by

$$\begin{aligned} A &= \pi b_1^2 \rho_0 v \\ &= \frac{2\pi GMR\rho_0}{v}. \end{aligned} \tag{3.5.3}$$

For a star like the Sun moving at $v = 20$ km/sec through a medium with $\rho_0 = 10^{-24}$ gm/cm^3 we have $A \approx 3 \times 10^7$gm/sec $\approx 5 \times 10^{-19}\ M_\odot$/yr, a negligible value.

The actual accretion rate of a collisional fluid is expected to be much larger than that given by (3.5.3), and was calculated by Hoyle

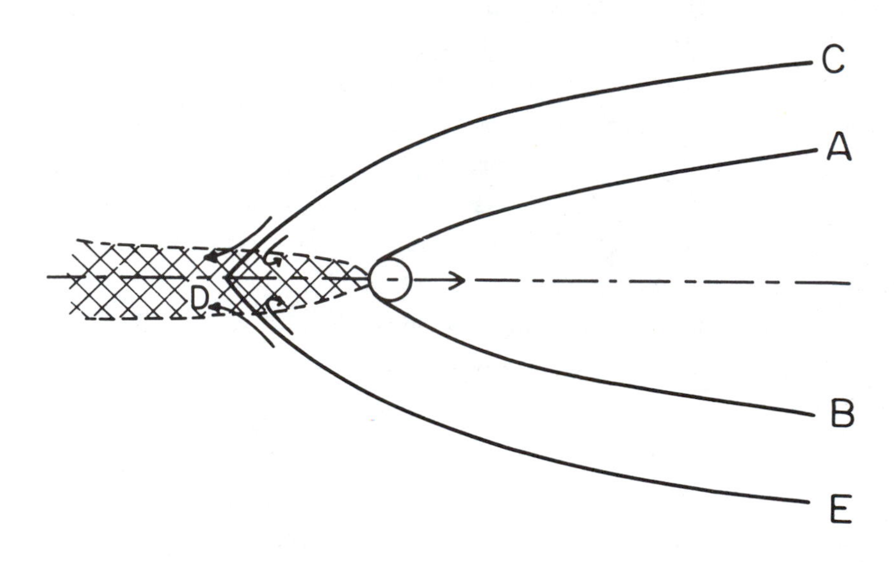

Figure 3.3. Accretion flow.

and Lyttleton (1939). Figure 3.3 shows a flow pattern stationary in the frame of the star, which is moving to the right. Fluid inside the nearly parabolic hyperbola AB (actually a hyperboloid of revolution whose axis is the star's direction of motion) falls directly onto the stellar surface, with the accretion rate (3.5.3). Material in paths with larger impact parameters, like CD and ED, converges on the axis of motion behind the star and undergoes particle collisions there. The accretion wake (qualitatively sketched as a cross-hatched area) contains particles which have collided; if mean free paths are short this region is sharply bounded by a thin shock.

All the fluid is on hyperbolic orbits about the star until collisions occur. They contribute to accretion in two ways. Collisions among atoms, ions with bound electrons, and molecules are often inelastic, leaving the collision products in excited states whose energy is usually promptly radiated, or produce dissociation and ionization; these processes reduce the particle kinetic energy. When collisions are frequent the matter follows the equations of hydrodynamics. Even if its total (kinetic plus internal) energy exceeds its gravitational binding energy, as it did before collisions began, there will be a flow in the accretion wake towards the attracting star. Matter close to the star will be in supersonic free-fall (this is an assumed boundary condition, which is justified if matter radiates and cools rapidly upon impact with the stellar surface, as is usually the case). No acoustic signal can propagate supersonically and reach the outer parts of the wake; there is no pressure support from the stellar surface, and the pressure and internal energy of the wake do not effectively oppose gravity. This is very different from a stellar wind, in which subsonically moving matter near the stellar surface supports and accelerates the outflow.

Roughly, we may say that matter in the wake whose kinetic energy alone is less than its gravitational energy will be accreted. The importance of the shock at the surface of the accretion wake is that it converts a portion of the kinetic energy of the matter to internal energy, and thus reduces its velocity below escape velocity.

To calculate the accretion rate we need to determine the impact parameter b for which the matter becomes bound after entering the accretion wake. On the wake axis the tangential component of velocity v_θ is zero (it drops by a large factor at the boundary shock, and is then reduced to zero by a pressure gradient within the wake). In order that the lost kinetic energy per unit mass equal that at infinity, v_θ of the matter entering the accretion wake must satisfy

$$\frac{1}{2}v_\theta^2 = \frac{1}{2}v^2. \tag{3.5.4}$$

The matter follows hyperbolic orbits, which are given by

$$r = \frac{b^2v^2/GM}{1 + e\cos\theta} \tag{3.5.5a}$$

where the eccentricity

$$e = \sqrt{\frac{b^2v^4}{G^2M^2} + 1}, \tag{3.5.5b}$$

and the angle θ is measured from the point of closest approach. The asymptote is at the angle $\theta_a = \cos^{-1}(-1/e)$, and the point D is at $\theta_D = \theta_a \pm \pi$, so that $\cos\theta_D = 1/e$. Then the distance r_D from D to the attracting star is

$$r_D = \frac{b^2v^2}{2GM}. \tag{3.5.6}$$

By conservation of angular momentum

$$r_D v_\theta = bv \tag{3.5.7}$$

so that at D

$$v_\theta = \frac{2GM}{bv}. \tag{3.5.8}$$

Now (3.5.4) is satisfied if

$$b = \frac{2GM}{v^2}, \tag{3.5.9}$$

implying an accretion rate

$$\begin{aligned} A &= \pi b^2 \rho_0 v \\ &= \frac{4\pi G^2 M^2 \rho_0}{v^3}. \end{aligned} \tag{3.5.10}$$

This result is larger than (3.5.3) by a factor $2GM/(v^2R)$, which is $\sim 10^3$ for typical stellar parameters. The accretion rate is still negligible under ordinary interstellar conditions ($\sim 10^{-15} M_\odot/\text{yr}$ for the Sun), but may be significant in other problems. The quadratic dependence of A on M is noteworthy; A is proportionally more important for very massive objects (large interstellar clouds, star clusters, and galaxies) than for ordinary stars.

The apparent divergence of A as $v \rightarrow 0$ is also noteworthy. We have so far assumed that the fluid is initially pressureless and has zero sound speed. Calculations by Bondi (1952) have shown that if $v = 0$ (3.5.10) gives the correct answer, if v is replaced by the sound speed c_s, and an uncertain (but ≈ 1) coefficient is introduced. A plausible interpolation formula between the $v \ll c_s$ and $v \gg c_s$ regimes is

$$A = \frac{4\pi G^2 M^2 \rho_0}{(v^2 + c_s^2)^{3/2}}. \tag{3.5.11}$$

Numerical calculations by Hunt (1971) have confirmed this expression. Perhaps his most important result is the applicability of (3.5.11) even when there are no radiative losses; this justifies the assertion than only kinetic, and not internal, energy contributes to the escape of fluid from the gravitational potential well.

For collisionless particles (3.5.3) should still be used; under typical interstellar conditions neutral atoms and molecules may be considered collisionless, and their accretion rate is very low. Thus the ionization state of the fluid determines its accretion rate. Careful calculations of interstellar accretion must consider photoionization by the radiation of the accreting object and the quantitative cross-sections for ion-neutral charge exchange and other collisional processes.

If the fluid is heterogeneous the accretion rate will be reduced below that of (3.5.11), because even a completely inelastic collision at D (in Figure 3.3) will not reduce v_θ to zero. Quantitative calculation requires a detailed description of the density distribution of the heterogeneous fluid, which is rarely available. The reduction of the accretion rate is not likely to be more than a factor ~ 2 unless density contrasts are high, because for b significantly less than that given by (3.5.9) v_θ at D substantially exceeds v, and loss of even half of v_θ

would leave the matter bound to the attracting object. Only in the case of small isolated particles moving in vacuum without dissipation is (3.5.3) correct.

3.6 Accretion Discs

In the preceding section we estimated the rate at which matter may be gravitationally captured by a star or compact object (degenerate dwarf, neutron star, or black hole). Its subsequent fate depends on its angular momentum about the source of gravitational attraction. If it has no angular momentum then matter will fall radially inward, and will soon either hit the surface of a star (releasing its gravitational binding energy as heat), or be swallowed by the event horizon of a black hole (in which case very little energy may be radiated).

Even a small amount of angular momentum drastically changes the flow. Matter with angular momentum per unit mass $\ell > \sqrt{2GMR}$ cannot fall directly onto the surface of a star of mass M and radius R. For neutron stars and black holes of comparable mass the maximum $\ell \approx 10^{16}$ cm^2/sec, while for degenerate dwarves it is $\approx 3 \times 10^{17}$ cm^2/sec. If a neutron star or a degenerate dwarf has a large magnetic field this may increase its effective size and maximum ℓ for direct accretion; this is discussed in **4.2.2**.

In the evolution of a binary system one star may expand beyond its limiting surface, known as the Roche lobe (or Roche limit), within which its gravity can confine matter. Outside this surface the gravity of its companion is sufficient to draw matter away from the expanding star, and toward the companion. This process, known as Roche lobe overflow, produces a smoothly flowing stream of matter from one star to the other. The rate of mass flow depends on the rate of evolution of the expanding star, the ratio of the stellar masses, and on the rates of angular momentum transfer and other relaxation processes within the binary; values over a large range are predicted (or observationally inferred) in different circumstances. The specific angular momentum ℓ of the transferred matter is comparable to that of the orbital motion, so that $\ell \gtrsim 3 \times 10^{18}$ cm^2/sec (depending on the stellar masses and the size of the orbit).

In accretion from a wind or other extended medium ℓ depends on the medium's heterogeneity. For typical interstellar accretion parameters ℓ may be as high as $\ell \sim bv\delta\rho/\rho \sim 10^{20}\delta\rho/\rho$ cm^2/sec, where $\delta\rho/\rho$ is the typical mean density heterogeneity averaged over the scale of the accretion impact parameter b (but see Livio 1986). In accretion from the high velocity wind of a binary companion star the spherical expansion of the wind guarantees a minimum $\delta\rho/\rho \sim 2b/D$ where D is the separation between the centers of the two stars (there may be additional sources of heterogeneity); for typical parameters the previous estimate gives $\ell \sim 10^{16}$ cm^2/sec. Under some (but not all) plausible circumstances sufficient angular momentum exists to prevent immediate accretion by the attracting star. We saw in **3.2** that the collapse of a protostar from an interstellar cloud is similarly limited by angular momentum.

Accretion discs were first studied for their application to the early Solar System; the pre-history of their study goes back to Laplace. Pringle (1981) presents a review with references to the extensive literature. The fundamental assumption in the theory of accretion discs is that matter radiates its internal energy much faster than it loses angular momentum. Examination of the rates of the many processes involved supports this. In fact, the central mystery of accretion disc physics is the mechanism by which angular momentum is transferred from one element of mass to another; ordinary viscous torques are insufficient. The lowest energy state of matter with given angular momentum is a circular Keplerian orbit; orbits with the same radius but different orientations collide, the energy of collision is radiated, and the matter settles into the circular orbit determined by its mean angular momentum.

Orbits with different radii need not be coplanar unless they are coupled by some dissipative process (models for some actual accretion discs in X-ray sources and SS433 require that they not be coplanar). Frictional torques are small because they are proportional to the viscosity. Gravitational torques are not dissipative and cannot directly enforce coplanarity, even if they are large; self-gravitational torques are small if disc masses are low. Despite these complications, simple models naturally assume a planar disc; deviations from flatness are considered as perturbations.

Self-gravity is usually neglected in disc models; this is believed (on somewhat uncertain grounds) to be justified in accretion discs arising from mass flows between the components of close binary stars, or from accretion onto stars and compact objects, but may well be wrong for protostellar discs. It is also assumed that pressure forces are small in comparison to the gravitational attraction of the central mass; this is justified by the assumption that radiation of energy is rapid. In the almost infinitesimally thin rings of Saturn, composed of solid particles, this approximation holds extraordinarily well. It is not empirically known how accurate this approximation is in stellar accretion discs; it may be incorrect, or at best rough. Finally, we assume Newtonian gravity; the thorough discussion by Novikov and Thorne (1972) includes relativistic effects.

The equations of structure of a fluid disc follow from these assumptions. On circular Keplerian orbits the velocity is entirely in the azimuthal (ϕ) direction:

$$v_\phi = \sqrt{GM/r}. \tag{3.6.1}$$

This is the fundamental approximation of thin disc theory. Because the disc matter is not exactly at zero temperature and zero pressure the disc will have a finite thickness h in the direction (z) along the angular momentum axis. If $h \ll r$ then the vertical and radial structure may be considered separately.

There is hydrostatic equilibrium between the z-components of the pressure gradient and the gravitational acceleration (by assumption, entirely that of the central mass). Expanding the gravitational force to lowest order in z/r, we obtain

$$\frac{\partial P}{\partial z} = -\rho \frac{GMz}{r^3}. \tag{3.6.2}$$

This force law is that of a harmonic oscillator. From this we obtain an estimate of h:

$$h \approx \sqrt{\frac{P}{\rho} \frac{r^3}{GM}} \approx \frac{c_s}{\Omega}, \tag{3.6.3}$$

where c_s is the sound speed and Ω is the angular frequency of Keplerian orbits about the central mass. Because of the square root in (3.6.3) h varies as the 1/2 power of the temperature, rather than being proportional to it, as is familiar from ordinary atmospheres; h should not be thought of as a scale height, for the density distribution is closer to Gaussian than exponential. A disc may, in addition, have a much thinner atmosphere whose scale height is proportional to its temperature.

Accretion discs are interesting (and observable) only if accretion actually takes place; that is, if matter flows through them. In order for matter to flow inward it must lose angular momentum, which flows outward. The net binding energy per unit mass of matter in Keplerian circular orbits is $-G^2M^2/(2\ell^2)$, so that in the state of lowest energy nearly all the disc mass has $\ell \to 0$ (it accumulates at zero radius, or on the surface of the attracting body), while all the angular momentum resides in an infinitesimal fraction of the disc mass in orbit at $r \to \infty$. The problem is to separate the mass from its angular momentum.

In a steady state in which there is a steady supply of matter with a finite $\ell > 0$, there is a continual flow of mass inward and of angular momentum outward. There must be a torque to extract angular momentum from the inward flowing mass. This torque is described as the consequence of a viscous stress, whatever its microscopic origin (which may be magnetic). The central problem of accretion disc theory is that it is not known, even to order of magnitude, how to calculate this stress; ordinary viscosity is certainly present but is much too small to explain the observed accretion rates. This problem is evaded by describing the stress by a parameter of unknown value.

The radial structure equations may be written in terms of quantities which are integrals over the vertical structure. The surface density is defined $\Sigma \equiv \int \rho \, dz$, and the mean viscous stress tensor $\langle t_{r\phi} \rangle \equiv \int t_{r\phi} \, dz/(2h)$; only the $r\phi$ component is important in a thin disc. In a steady state disc, conservation of mass may be expressed

$$-2\pi r v_r \Sigma = \dot{M}, \tag{3.6.4}$$

where $\dot{M}$ is the mass flow rate and v_r the (mass-averaged) mean radial component of velocity. The conservation of angular momentum takes the form

$$\dot{M}\sqrt{GMr} = (2\pi r)(2h\langle t_{r\phi} \rangle r) + \dot{J}. \tag{3.6.5}$$

The left hand side is the rate at which mass flow carries angular momentum inward across a cylinder of radius r. In steady state the disc inside this radius does not accumulate angular momentum, so this flow equals the viscous torque the disc inside r exerts on that outside r plus the rate $\dot{J}$ at which angular momentum is taken up by the central attracting mass.

$\dot{J}$ depends on the boundary conditions at the central object, which depend on its mass, angular momentum, and magnetic field, and on the radius r_I of the inner edge of the disc. Like other uncertainties, it is readily parametrized:

$$\dot{J} = \varsigma \dot{M} \sqrt{GMr_I}. \tag{3.6.6}$$

The parameter ς is expected to be in the range $0 \leq \varsigma \leq 1$; $\varsigma < 0$ would correspond to a rapidly rotating central object supplying angular momentum to the disc; such a process may occur (and may be necessary to explain the very long spin periods observed for some neutron stars), but probably only in a disc in which the central object does not accrete at all, and all matter flows outwards. In most cases $\varsigma = 1$ is probably a fair assumption, equivalent to assuming that the central object acquires the angular momentum of the accreted matter along with its mass.

From (3.6.5) and (3.6.6) we obtain the mean stress

$$2h\langle t_{r\phi}\rangle = \frac{\dot{M}\sqrt{GMr}}{2\pi r^2}\left(1 - \varsigma\sqrt{\frac{r_I}{r}}\right). \tag{3.6.7}$$

In order to find the heat released we need the rate-of-strain tensor, whose $r\phi$ component is given, for Keplerian orbits and $v_r \ll v_\phi$, by

$$\begin{aligned}\sigma_{r\phi} &= \frac{1}{2}\left(\frac{1}{r}\frac{\partial v_r}{\partial \phi} + \frac{\partial v_\phi}{\partial r} - \frac{v_\phi}{r}\right)\\ &= -\frac{3}{4}\sqrt{\frac{GM}{r^3}}.\end{aligned} \tag{3.6.8}$$

The rate of viscous heating per unit volume is given by

$$\epsilon = -2t_{r\phi}\sigma_{r\phi}, \tag{3.6.9}$$

and the total power released per unit area $2H \equiv \int \epsilon \, dz$ is

$$2H = \frac{3GM\dot{M}}{4\pi r^3}\left(1 - \varsigma\sqrt{\frac{r_I}{r}}\right). \tag{3.6.10}$$

Note that for $r_I \ll r$ or $\varsigma = 0$ (3.6.10) is three times the rate at which gravitational energy is released by the progression of matter to more tightly bound orbits; the source of the extra energy is work done by the viscous stress. Because a disc has two surfaces, each element of surface radiates a power per unit area H.

The total disc luminosity L is found, if the disc extends to very large radii:

$$\begin{aligned} L &= \int_{r_I}^{\infty} 2H\ 2\pi r\ dr \\ &= \left(\frac{3}{2} - \varsigma\right) \frac{GM\dot{M}}{r_I}. \end{aligned} \tag{3.6.11}$$

This is the sum of two terms. $GM\dot{M}/(2r_I)$ is the binding energy of the Keplerian orbit at radius r_I, while $(1-\varsigma)GM\dot{M}/r_I$ represents work done by the central object on the disc. The kinetic energy of the Keplerian orbit at r_I may be dissipated and radiated in a narrow boundary layer at the surface of an accreting star, but because matter in this boundary layer does not satisfy (3.6.1) disc theory is inapplicable there.

At each point on its surface a disc radiates the flux H. Most of the energy is released at radii of order but not very close to r_I. For $\varsigma = 1$ half is released between r_I and $4r_I$, and an additional 40% inside $26r_I$. As a rough approximation we assume the disc surface radiates as a black body, in which case the disc surface effective temperature is given by

$$\begin{aligned} T_e &= \left(\frac{H}{\sigma_{SB}}\right)^{1/4} \\ &= \left[\frac{3GM\dot{M}}{8\pi r^3 \sigma_{SB}}\left(1 - \varsigma\sqrt{\frac{r_I}{r}}\right)\right]^{1/4} \\ &\sim 5 \times 10^7\ {}^{\circ}\mathrm{K}\ \dot{M}_{17}^{1/4} \left(\frac{M}{M_\odot}\right)^{-1/2} \left(\frac{GM}{rc^2}\right)^{3/4} \left(1 - \varsigma\sqrt{\frac{r_I}{r}}\right)^{1/4}, \end{aligned} \tag{3.6.12}$$

where $\dot{M}_{17} \equiv \dot{M}/(10^{17}\ \mathrm{gm/sec})$. For typical X-ray source parameters most of the power is radiated in soft X-rays ($h\nu \sim 3$ KeV). For parameters appropriate to the supermassive black holes suggested for quasars, the disc radiation would appear in the ultraviolet.

If $\varsigma = 1$ the term proportional to it has the effect of moving the peak H and T_e to $r = \frac{49}{36} r_I$; the peak H and T_e are respectively .057 and .49 of their values for the case $\varsigma = 0$. If there is a thin boundary layer it will have H and T_e much higher than the disc values given by (3.6.10) and (3.6.12).

In simple disc models the spectrum is given by an integral over black body spectra, with T_e given by (3.6.12). At low frequencies ($h\nu \ll k_B T$, where T is an inner disc temperature), a power law is predicted. Because of the exponential cutoff of a black body spectrum, the Planck function (1.7.13) may be very roughly approximated as $\propto \nu^2 T$ for $h\nu < 3k_B T$, and zero for $h\nu > k_B T$. Then the integrated spectrum I_ν is estimated, to order of magnitude, by

$$\begin{aligned} I_\nu &= \int_{r_I}^{\infty} B_\nu(T(r)) 2\pi r \, dr \\ &\propto \int_{r_I}^{r_{max}} \nu^2 T(r) r \, dr, \end{aligned} \tag{3.6.13}$$

where $h\nu = 3k_B T(r_{max})$ defines r_{max}. For $r_{max} \gg r_I$ we have $T(r) \propto r^{-3/4}$ over most of the range of integration, so that

$$\begin{aligned} I_\nu &\propto \nu^2 \int_{r_I}^{r_{max}} r^{1/4} \, dr \\ &\propto \nu^2 r_{max}^{5/4}. \end{aligned} \tag{3.6.14}$$

But $r_{max} \propto \nu^{-4/3}$, so that a power law is predicted

$$I_\nu \propto \nu^{1/3}. \tag{3.6.15}$$

Unfortunately, (3.6.12) and (3.6.15) are not supported by any data. There are few astronomical objects in which the continuum radiation from an accretion disc can be unambiguously identified. One likely case is the black hole candidate Cygnus X-1 (a black hole has no stellar surface as an alternative source of radiation). Its X-ray spectrum requires temperatures $10 - 100$ times higher than predicted. The most plausible explanation is that most of the energy release occurs in very hot optically thin regions, in disagreement with the assumption of a black body radiator. There is no good way to predict the properties of these regions, and Cygnus X-1 shows very different spectra at different times, so that any single model is at least sometimes wrong. Stars known as dwarf novae have outbursts in which most of their radiation is believed to come from an accretion disc. The inferred temperatures are at least approximately consistent with (3.6.12), but data are not available over a wide enough range of frequency to test (3.6.15). Because these are transient outbursts, steady state disc models should be inapplicable.

We were able to evade the question of disc viscosity by obtaining results parametrized by the accretion rate. $\dot{M}$ is directly related to L and other observable quantities, and may be estimated, even though the orders of magnitude of Σ and v_r are unknown. If assumptions are made about the viscosity these quantities may be calculated, along with the detailed structure of the disc; the results are no more certain than the assumptions. It is usually found that $h \ll r$, typically by one to two orders of magnitude, and that h is nearly proportional to r in a given disc, except in hot inner regions; these conclusions are nearly independent of the numerical values assumed for the viscosity or v_r. They are consistent with the assumption of a thin disc, but are not confirmed by any observational data. Uncertain arguments and models of SS433 suggest that in the outer regions of its disc $h/r \sim 0.5$.

It is possible to derive an interesting result for the thickness of a luminous disc. Begin with the expression (3.6.3) for h, and substitute

$P = P_r/(1-\beta)$, where $\beta \equiv P_g/P$ is taken to be constant:

$$h \approx \sqrt{\frac{P_r}{\rho(1-\beta)} \frac{r^3}{GM}}. \tag{3.6.16}$$

Now use (1.7.19) and (3.6.10):

$$\begin{aligned} H &= \frac{c}{\kappa\rho}\frac{dP_r}{dz} \\ \frac{3GM\dot{M}}{8\pi r^3}\left(1 - \varsigma\sqrt{\frac{r_I}{r}}\right) &\approx \frac{cP_r}{\kappa\rho h}, \end{aligned} \tag{3.6.17}$$

Substituting this result for P_r into (3.6.16) leads to

$$h \approx \frac{3\kappa\dot{M}}{8\pi c(1-\beta)}\left(1 - \varsigma\sqrt{\frac{r_I}{r}}\right). \tag{3.6.18}$$

Define the energetic efficiency of accretion ε by $L = \dot{M}c^2\varepsilon$, and the characteristic accretion rate $\dot{M}_E$ in terms of the Eddington limiting luminosity L_E (1.11.6):

$$\begin{aligned} \dot{M}_E &\equiv \frac{L_E}{c^2\varepsilon} \\ &= \frac{4\pi GM}{\kappa c\varepsilon}. \end{aligned} \tag{3.6.19}$$

From (3.6.11)

$$\varepsilon = \left(\frac{3}{2} - \varsigma\right)\frac{GM}{r_I c^2}. \tag{3.6.20}$$

(3.6.18) may be rewritten, if $\beta \ll 1$ (as is found to be the case in the inner regions of luminous discs)

$$h \approx r_I\left(\frac{3}{3-2\varsigma}\right)\left(\frac{\dot{M}}{\dot{M}_E}\right)\left(1 - \varsigma\sqrt{\frac{r_I}{r}}\right). \tag{3.6.21}$$

Then h varies only slowly with r. The maximum value of h/r is found to be

$$\begin{aligned} &\left(\frac{4}{27\varsigma^2 - 18\varsigma^3}\right)\left(\frac{\dot{M}}{\dot{M}_E}\right) \quad \text{at} \quad r = \frac{9}{4}\varsigma^2 r_I \qquad \text{if } \varsigma \geq \frac{2}{3}, \\ &\left(\frac{3 - 3\varsigma}{3 - 2\varsigma}\right)\left(\frac{\dot{M}}{\dot{M}_E}\right) \quad \text{at} \quad r = r_I \qquad \text{if } \varsigma \leq \frac{2}{3}. \end{aligned} \tag{3.6.22}$$

These results imply the existence of a characteristic disc accretion rate and radiative luminosity, related to the limiting radiative luminosity of stars. If $\dot{M} \gtrsim \dot{M}_E$ the inner regions of a disc have $h \gtrsim r$, and thin disc theory is inapplicable. Disc accretion at such a high rate may qualitatively resemble radial accretion, which depends on the nature of the central object. The energy released by accretion is trapped as radiation within the accreting matter (**3.7**). If the attracting object has a surface, then an extended envelope close to hydrostatic equilibrium will rapidly accumulate, while in accretion onto a black hole the radiation is swept into the black hole along with the matter (Eggum, *et al.* 1985). Even if $\dot{M} \gg \dot{M}_E$, the emergent luminosity $L \lesssim L_E$.

Thin discs and non-rotating stars may be thought of as two opposite limits of a continuum. In a star the force of gravity is balanced by a pressure gradient, while in a thin disc gravity is balanced by angular momentum. Most astronomical objects, other than those in which flows are chaotic, may be approximately described by one limit or the other. For this reason the understanding of discs may be as important to astrophysics as that of stars.

Intermediate configurations may exist; near the stellar end of the continuum they may be considered rotationally flattened stars, while near the thin disc end they are discs thickened by their internal pressure. In a boundary layer between a thin disc and a slowly rotating star there is a continuous transition between these two limits.

The theory of discs is in a much more primitive state than that of stars, because one essential constitutive relation is not understood, their rate ϵ of viscous heating. This resembles the problem of stellar structure prior to the development of nuclear physics in the 1930's. We may be worse off than this, because so few direct observations of discs are possible. What little data exist (for example, for discs around likely black holes like Cygnus X-1) indicates that real discs are not steady objects radiating from optically thick photospheres (as the theory assumes), but that they are wildly variable, release much of their energy in optically thin regions, and may have important nonthermal processes. It may be appropriate to compare our present understanding of discs to Galileo's understanding of sunspots and solar activity.

3.7 Radial Infall

There are at least three reasons for considering radial accretion flows, in which matter with negligible angular momentum falls radially onto a star, compact object, or black hole. In some circumstances matter with very little angular momentum may be accreted, and infall is nearly radial. This need not imply that the flow is spherically symmetric; for example, flow from the accretion wake shown in Figure 3.3 may be radially inward, but is concentrated onto a small part of the stellar surface. The second reason is that if the accreting object has a large magnetic field this field will force the infall of conducting matter near it to flow along the field lines, and in particular along those field lines which extend to great distances. In this case the flow near the star will be nearly radial, and may therefore closely approximate radial accretion. This is believed to be a good description of accretion onto many magnetized neutron stars and degenerate dwarves. The final reason is that radial accretion is a relatively simple and calculable problem, which stands at the opposite extreme from that of an angular momentum-dominated accretion disc. Even were radial accretion never realized, its study would illuminate the range of accretion flows possible in more complex circumstances.

In free fall from infinity the infall velocity v_r is

$$v_r = -\sqrt{\frac{2GM}{r}}. \tag{3.7.1}$$

For an accretion rate $\dot{M}$ the density is

$$\rho = \frac{\dot{M}}{|v_r| r^2 \Omega}, \tag{3.7.2}$$

where the flow has been assumed uniform over a solid angle Ω. In a spherically symmetric flow $\Omega = 4\pi$, but in other cases Ω may be much less. If the flow is guided by a dipole magnetic field the solid angle is determined by the size of a bundle of field lines. Because $B \propto r^{-3}$ the cross-section of such a bundle is $\propto r^3$, and $\Omega \propto r$.

The most important parameter describing radial accretion is $\dot{M}$. If $\dot{M}$ is small then the accretion luminosity

$$L = \dot{M} c^2 \varepsilon \tag{3.7.3}$$

is small, the outward force F_{rad} of radiation pressure on the infalling matter is negligible, and (3.7.1) and (3.7.2) are applicable. In spherical geometry

$$F_{rad} = \frac{L\kappa}{4\pi r^2 c}, \tag{3.7.4}$$

where κ is the opacity. Properly, κ should be frequency-averaged, but in most accretion flows the opacity is almost entirely frequency-independent electron scattering, so κ may be taken to be κ_{es}. The ratio of F_{rad} to the inward force of gravity is

$$\begin{aligned}\frac{F_{rad}}{F_{grav}} &= \frac{L\kappa}{4\pi cGM} \\ &= \frac{L}{L_E},\end{aligned} \tag{3.7.5}$$

where the Eddington luminosity $L_E = 4\pi cGM/\kappa$ was defined for stellar interiors in (1.11.6). If $L \ll L_E$ the influence of radiation pressure may be ignored. If $L > L_E$ the net force on matter is directed outward! It is also possible to define the same Eddington accretion rate $\dot{M}_E \equiv L_E/(c^2\varepsilon)$, where $c^2\varepsilon$ is the energy release per gram, as we did for disc accretion in (3.6.19).

It is clear that L_E is an upper bound on the luminosity emerging from an accreting object, just as it limits the radiative luminosity of a star in hydrostatic equilibrium, and the applicability of thin disc models. It is less obvious what happens if $\dot{M} > \dot{M}_E$; that is, if an accretion flow "tries" to exceed L_E. For example, an external source of mass may supply it at a rate exceeding $\dot{M}_E$. In stellar interiors in hydrostatic equilibrium luminosities in excess of L_E can be carried by convection, but this is inapplicable to accretion flows, which are very far from hydrostatic.

If the accreted matter falls freely until it hits the stellar surface, the optical depth τ along a path radially outward from the surface to infinity is

$$\begin{aligned}\tau &= \int_R^\infty \kappa\rho \, dr \\ &= \frac{\kappa\dot{M}}{2\pi\sqrt{2GMR}}.\end{aligned} \tag{3.7.6}$$

We have assumed a spherically symmetric inflow. This may be rewritten, using the definition of $\dot{M}_E$ and the Newtonian value of $\varepsilon = GM/(Rc^2)$. The result, written in terms of the escape velocity

$v_{esc} = \sqrt{2GM/R} < c$, is

$$\begin{aligned}\tau &= \left(\frac{\dot{M}}{\dot{M}_E}\right)\sqrt{\frac{2GM}{Rc^2}}\frac{1}{\varepsilon} \\ &= 2\left(\frac{\dot{M}}{\dot{M}_E}\right)\frac{c}{v_{esc}}.\end{aligned} \tag{3.7.7}$$

As $\dot{M}$ approaches $\dot{M}_E$ the flow becomes optically thick.

It is possible to describe crudely the flow of radiative energy through optically thick matter by a diffusion velocity v_{diff}, using (1.7.15b):

$$\begin{aligned}v_{diff} &\equiv \frac{H}{\mathcal{E}_{rad}} \\ &\sim \frac{c}{3\tau};\end{aligned} \tag{3.7.8}$$

this result is valid only if $\tau \gtrsim 1$. Then we may compare the rate at which free-falling matter flows inward to the rate at which radiation diffuses outward through the matter:

$$\frac{v_{diff}}{v_{esc}} \sim \frac{1}{6}\left(\frac{\dot{M}}{\dot{M}_E}\right)^{-1}. \tag{3.7.9}$$

Before $\dot{M}$ can reach $\dot{M}_E$ the matter, optically thick to the radiation, effectively traps it and sweeps it inward, preventing its escape. Once $\dot{M} \gtrsim \dot{M}_E$, the approximations used in deriving (3.7.9) become invalid. The radiation trapped deep within the flow builds up to a higher energy density, and slows the infall. Because radiation is trapped within the inner part of the flow, in its outer part F_{rad} is smaller than (3.7.4) would imply, and there matter may continue to accrete; it is not blown away by radiation pressure, as might be imagined.

In the limit $\dot{M} \gg \dot{M}_E$ the radiative transport of energy is a small effect on infall time scales, and the radiation and matter should be thought of as inseparable components of a single fluid, with the combined equation of state (1.4.5). When the fluid reaches the stellar surface a strong shock forms, which reduces its velocity and increases its density. The compression ratio (3.3.17) is not large, (an isothermal equation of state is impossible because it would require efficient radiative losses, and we have seen that radiation is trapped within the matter). The accreted matter very rapidly builds up an opaque and voluminous (but low mass) envelope on the star, nearly in hydrostatic equilibrium; the shock marks the outer boundary of the envelope. The accreting compact object is very soon lost from view beneath this envelope, which roughly resembles that of a supergiant star.

As an approximation to the structure of the envelope we assume that once matter passes through a strong shock at radius r it remains at r, having converted all its kinetic energy to internal energy; this ignores its small residual velocity and kinetic energy. Its internal energy $\mathcal{E}$ is given by

$$\mathcal{E} = \frac{GM\rho}{r}. \tag{3.7.10}$$

Then, substituting $P = (\gamma - 1)\mathcal{E}$ (**1.9**) into the equation of hydrostatic equilibrium

$$\frac{\partial P}{\partial r} = -\frac{GM\rho}{r^2} \tag{3.7.11}$$

yields the solution

$$\rho \propto r^{(\gamma-2)/(\gamma-1)}. \tag{3.7.12}$$

Under most conditions of interest radiation pressure is dominant and $\gamma = 4/3$, so that $\rho \propto r^{-2}$; this is borne out by radiation-hydrodynamic calculations (Klein, *et al.* 1980, Burger and Katz 1983).

This result should be compared to the structure of a polytropic envelope. Substituting $P \propto \rho^\gamma$ into (3.7.11) yields

$$\rho \propto r^{-1/(\gamma-1)}. \tag{3.7.13}$$

This is a steeper variation of ρ with r than (3.7.12), implying that the envelope structure (3.7.12) is unstable against convection. If it remains in hydrostatic equilibrium for a long enough time turbulent convection will make its structure approach that of (3.7.13).

In accretion onto a black hole there is no surface to support an envelope, and free fall is likely at any accretion rate.

3.8 Jets

A number of diverse astronomical objects contain two oppositely directed streams of matter flowing outward from a common source, called jets (Ferrari and Pacholczyk 1983). This phenomenon is found in quasars and galaxies with active nuclei (Begelman, *et al.* 1984), in the peculiar binary star SS433 (Margon 1984, Katz 1986), in the dense dusty gas clouds surrounding protostars called bipolar nebulae (Lada 1985), and possibly in other kinds of objects. The degree of collimation of the jets, the nature of the accelerated material, its speed, and the symmetry between the two opposed jets all vary between classes of jets, and to a significant extent within a given class.

Extragalactic jets range from tightly collimated ($\sim 1°$) filaments to broad diffuse blobs. They originate in galactic nuclei, and often terminate in extended lobes of radio emission, the well known "double radio sources." Often one jet is much brighter than the other. They may be broken into bright knots separated by regions in which the jet is faint or undetectable. The observed radiation is the synchrotron emission of very energetic electrons in weak magnetic fields, observed at radio wavelengths (and occasionally at visible wavelengths). There is little evidence for thermal plasma in the jets, although it could be present and even dominate the jet energy flow without being observable.

In SS433 the observed jets are composed of singly ionized thermal plasma, with a temperature $T \sim 10^4$ °K, radiating in the spectral lines of hydrogen and helium. This matter is moving at a speed of 78,000 km/sec in jets which are collimated to a few degrees. Aside from the large Doppler shifts, the spectrum resembles that of ordinary ionized stellar or interstellar material. A very small fraction of the jet energy is converted to the energy of relativistic electrons, which produce observable radio jets by synchrotron emission. The jets of SS433 do not have fixed orientations, but rather trace out the surface of a cone of half-angle 20° with a period of 164 days. This motion (which has additional complications) provides important clues to the nature of SS433 and to the jet acceleration mechanism.

The jets in the bipolar nebulae are usually poorly collimated, generally resembling diffuse clouds. They consist of molecular gas with velocities $\sim$ 100 km/sec and temperatures $\lesssim$ 100°K. Nonthermal phenomena (relativistic particles and synchrotron emission) appear to be absent.

These phenomena are very diverse, but have in common the geometry of two opposite jets emanating from a common source. The origin of this geometry must be very general, for it is found for relativistic particles, thermal plasmas, and neutral gases, in flows ranging over several orders of magnitude in size, power, energy density, velocity, sound speed, temperature, optical depth, particle mean free path (and almost any other parameter one cares to compute). Angular momentum has the required symmetry, for it determines a preferred axis through a center of mass. As discussed in **3.6**, angular momentum is expected to dominate the dynamics of matter drawn from a large volume by a small attracting object, or exchanged between two stars in a binary system. Matter with a significant amount of angular momentum naturally forms a disc around the center of gravitational attraction. Its axis then defines two opposite (but equivalent) preferred directions, which may be identified with the jet directions.

Both accretion discs and jets are consequences of the frequent importance of angular momentum in astrophysics, just as objects in which it is unimportant are spherical.

The symmetry of jet geometry is easy to explain. It is more difficult to understand the mechanisms of jet production and collimation, and why jets have their observed properties. If a bubble of buoyant, high entropy fluid is produced within a star or an accretion disc it will rise to the surface, with its buoyancy force directed parallel to the hydrostatic equilibrium pressure gradient and opposite to the local effective (including the centripetal potential) acceleration of gravity. In a spherical star this force is everywhere radially outward; in a rotationally flattened star it is preferentially directed toward the poles; in a thin accretion disc or rotationally flattened gas cloud it is parallel (or anti-parallel) to the rotational axis. The rise of such buoyant fluid is analogous to the motions of ordinary stellar convection, but we now consider material whose density is very much less than, rather than close to, that of its mean surroundings. Possible examples include a magnetized fluid of relativistic particles produced by pulsars or other nonthermal processes, or hot thermal fluid heated by strong shocks. In the course of its rise the buoyant fluid may mix with or entrain the surrounding denser fluid, a process which should be described by an (unknown) parameter analogous to the mixing length of ordinary convection.

Once the buoyant fluid breaches the surface of the disc or gas cloud it will expand; if its internal energy exceeds the gravitational binding energy it will escape freely. This flow may be either steady or intermittent, taking the form of discrete bubbles or persistent channel flow, depending on whether the supply of fluid is steady or intermittent, and on values of other hydrodynamic parameters; Norman, *et al.* (1981) present calculations. The expansion is preferentially normal to the surface (or to surfaces of constant pressure), and along the angular momentum axis, thus forming jets. Their degree of collimation depends on the detailed hydrodynamics of the flow. This is crudely analogous to the behavior of gas bubbles produced by underwater explosions, whose breaching can produce prominent vertical spikes of spray.

The expansion of the escaping buoyant fluid converts its internal energy to the kinetic energy of bulk motion (a process known as adiabatic loss of internal energy, because the expansion is usually taken to be adiabatic). Encounter with a surrounding medium or magnetic field may re-randomize the particle velocities in a collisional or (more likely, if densities are very low) collisionless shock or instability, possibly at very great distances from the source.

In an alternative class of models matter is heated or particles are accelerated at or above the surface of an accretion disc. The hot or energetic material need not rise through denser fluid. If its internal energy exceeds the gravitational binding energy it will escape, with its motion preferentially along the angular momentum axis. The source of its internal energy may be nonthermal particle acceleration, or a thermal process; Eggum, *et al.* (1985) report calculations of thermal jets produced by the radiation pressure of accretion discs with $\dot{M} > \dot{M}_E$. The collimation of the jets is determined by the detailed hydrodynamics of the flow and by the geometry of the accretion disc; the dense disc provides a boundary condition on the flow, roughly akin to a rocket nozzle, as well as a source of radiative acceleration (**1.15**).

These processes depend on complex, often turbulent, hydrodynamics with uncertain boundary conditions. In the case of extragalactic jets consisting of nonthermal particles the poorly understood processes of nonthermal particle acceleration are central. It is possible that all the ingredients of complete models of jets are qualitatively known, but turning them into a quantitative predictive (or even explanatory) tool remains a difficult problem.

After a jet leaves the vicinity of its source it will interact with any surrounding medium. This interaction may be complex; in particular, nonthermal processes may accelerate or scatter energetic electrons which produce radio (and occasionally visible) synchrotron radiation. A very rough model based only on momentum balance is capable of describing the rate at which a jet carves a cavity in the medium, and therefore relates the apparent length of a jet to the period over which it has been produced and to the physical parameters of the jet and of the medium in which it propagates.

The flow is sketched in Figure 3.4. A collimated jet of density ρ_1 travelling at speed u enters a semi-infinite medium at rest. The collision between the jet and the medium slows and deflects the jet, and pushes the medium aside; the jet creates for itself a cavity. At the tip of the jet fresh material interacts with the medium. This interaction region I moves to the right at a speed U. It is possible to calculate U without considering the complex hydrodynamics of the interaction region.

In a frame moving with I the flow is stationary, at least when averaged over any turbulence which may be present. The jet impinges on I from the left at a speed (assuming all velocities to be nonrelativistic) $u - U$, and the medium flows towards I from the right at a speed U. Because there are no other sources of momentum and I contains negligible mass (and is not accelerated), these two momentum fluxes balance

$$\rho_1(u - U)^2 = \rho_2 U^2. \tag{3.8.1}$$

Simple algebra gives U:

$$U = \frac{u}{1 + \sqrt{\rho_2/\rho_1}}. \tag{3.8.2}$$

In a time t the jet penetrates to a depth $D = Ut$. Equation (3.8.2) has been verified empirically for nearly incompressible fluid flow, if

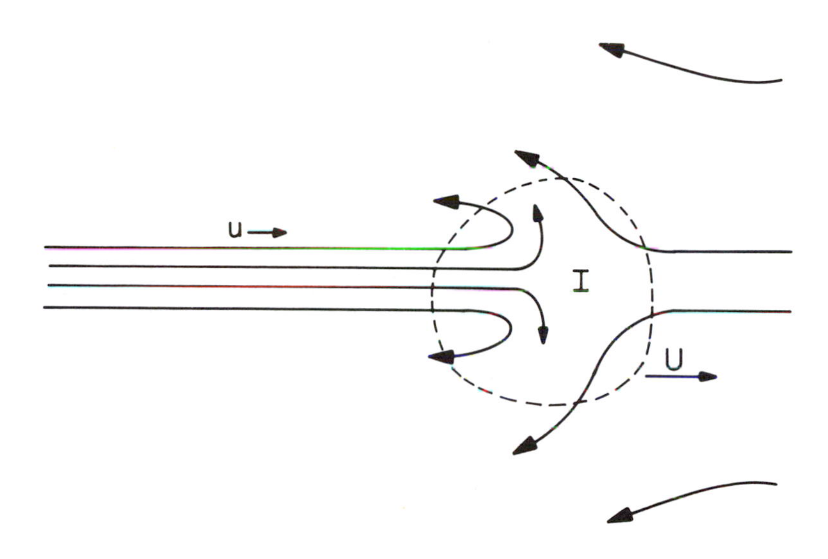

Figure 3.4. Jet propagation.

there is little entrainment of the medium by the jet before it reaches I. In this case the description of the complex flow pattern at I by a simple equation of momentum balance appears to be valid. Real astrophysical jets are much more complex, and the validity of this simple model of their hydrodynamics is uncertain. For example, because the flow is supersonic strong shocks form, and will tend both to disrupt the jet before it reaches region I and to evacuate the medium.

This calculation is readily generalized to relativistic velocities. The results are not so simple, except in limiting cases. Define ρ_1 to be the jet's rest mass density measured in the laboratory frame.

Then

$$U \approx \begin{cases} \sqrt{\dfrac{\rho_1/\rho_2}{1-u^2}}, & \text{if } \dfrac{\rho_1}{\rho_2} \ll 1-u^2; \\ u, & \text{if } \dfrac{\rho_1}{\rho_2} \gg 1-u^2. \end{cases} \tag{3.8.3}$$

Each of these limits may be interpreted as a consequence of momentum balance.

3.9 Magnetohydrodynamics

3.9.1 Equations If a fluid is an electrical conductor, as is most of the matter dealt with in astronomy, currents may flow in it and produce magnetic fields. These fields exert forces on the fluid through the currents it carries. Magnetic fields are observed in many astronomical objects. Even when they are not observed they may be present with magnitudes too small to permit direct observation, but large enough to produce other interesting consequences. For example, magnetic fields may carry significant torques (**3.6**), and time-dependent fields may lead to the acceleration of energetic particles (**2.5**).

The generalization of the laws of hydrodynamics to a conducting fluid is called magnetohydrodynamics. In each case the fluid equations are derived from kinetic equations (**2.1**) by assuming that the particle distribution functions are very close to those of thermodynamic equilibrium. It there are at least two species of particles with opposite signs of charge, the mean charge density is nearly zero, displacement currents are neglected, and the equations of electrodynamics are used to calculate the electromagnetic fields and the forces on the charged particles, then the equations of magnetohydrodynamics are obtained. The neglect of the displacement current and of the net charge density mean that the high frequency phenomena of electrostatic plasma waves and transverse electromagnetic waves are excluded.

The equations of magnetohydrodynamics are inapplicable if particle mean free paths and relaxation times are long, as is generally the case for energetic particles in low density media, because then the assumption of thermodynamic equilibrium (the complete description of the fluid by its pressure, density, velocity, and magnetic field, all of which are functions of space and time) fails. Even when strictly inapplicable, the magnetohydrodynamic equations may be a useful qualitative description of the fluid, although they will not describe some real phenomena. They may be a much better description of a magnetized (but nonequilibrium) fluid than the equations of hydrodynamics are of a similar unmagnetized fluid, for the gyration of energetic charged particles around the magnetic field lines may in effect shorten their mean free path to their gyroradius, which is generally very small.

Following Jackson (1975), the equation of conservation of mass is (3.1.1):

$$\frac{\partial \rho}{\partial t} + \nabla \cdot (\rho \vec{u}) = 0. \tag{3.9.1}$$

The equation of momentum conservation is essentially the same as (3.1.5), but we write the magnetic (Lorentz) force explicitly:

$$\frac{\partial \vec{u}}{\partial t} + (\vec{u} \cdot \nabla)\vec{u} = -\frac{1}{\rho}\nabla P + \frac{\vec{J} \times \vec{B}}{\rho c} + \frac{\vec{F}}{\rho}, \tag{3.9.2}$$

where $\vec{F}$ describes any other body forces, such as gravity and viscous stress. Neglecting the charge density and the displacement current, the electromagnetic field equations become

$$\nabla \times \vec{E} = -\frac{1}{c}\frac{\partial \vec{B}}{\partial t} \tag{3.9.3}$$

$$\nabla \times \vec{B} = \frac{4\pi}{c}\vec{J} \tag{3.9.4}$$

$$\nabla \cdot \vec{E} = 0 \tag{3.9.5}$$

$$\nabla \cdot \vec{B} = 0. \tag{3.9.6}$$

In addition to the relation (3.1.18) which determines the pressure P, another constitutive relation is needed to determine the current density $\vec{J}$. The simplest possible relation is the elementary Ohm's law, which ignores the Hall effect and a variety of other thermomagnetic and galvanomagnetic phenomena. In the rest frame of an element of fluid this law takes the form

$$\vec{J}' = \sigma \vec{E}', \tag{3.9.7}$$

where the primes denote the fluid frame, and σ is a scalar conductivity. Because we are neglecting charge densities, and assuming that all velocities are nonrelativistic, the current density in any frame $\vec{J} = \vec{J}'$. The electric field $\vec{E}'$ may be expressed in terms of the fields $\vec{E}$ and $\vec{B}$ in an inertial reference frame, so that

$$\vec{J} = \sigma(\vec{E} + \frac{\vec{u}}{c} \times \vec{B}). \tag{3.9.8}$$

In real astronomical problems the validity of (3.9.7) and (3.9.8) may be questionable because they do not correctly describe the response of collisionless particles to applied fields. They are valid only in the limit in which particle distribution functions undergo rapid collisional relaxation. There is no simple remedy for this problem, and it is likely to be serious whenever the fluid contains energetic particles.

In good conductors the electric fields are very small, and usually not directly measurable. They are still important, because without them no current would flow and there would be no magnetic field. It is usually of more interest to examine the behavior of the magnetic field, which is often large and measurable, using equations in which the electric field does not appear explicitly. The electric field may be eliminated from (3.9.3) by use of (3.9.8), yielding

$$\frac{\partial \vec{B}}{\partial t} = \nabla \times (\vec{u} \times \vec{B}) - \frac{c}{\sigma} \nabla \times \vec{J}. \tag{3.9.9}$$

The current density is also hard to measure directly, but may be eliminated by taking the curl of (3.9.4), using an elementary vector identity, and substituting into (3.9.9):

$$\frac{\partial \vec{B}}{\partial t} = \nabla \times (\vec{u} \times \vec{B}) + \frac{c^2}{4\pi\sigma} \nabla^2 \vec{B}. \tag{3.9.10}$$

3.9.2 Resistive Decay If the fluid is everywhere at rest ($\vec{u} = 0$) then (3.9.10) becomes a diffusion equation, and the magnetic field gradually decays to a uniform value. This value is set by the boundary conditions, and is generally zero for an isolated object. The magnetic energy appears as resistive heating of the matter. In a body or region of size R the characteristic decay time t_{md} is approximately

$$t_{md} \sim \frac{4\pi\sigma R^2}{c^2}. \tag{3.9.11}$$

For an ionized nondegenerate hydrogen plasma

$$\sigma \approx \frac{1.4 \times 10^8\ T^{3/2}}{\ln \Lambda}\ \mathrm{sec}^{-1}, \tag{3.9.12}$$

where $\ln \Lambda$ is the Coulomb logarithm (2.2.10) (Spitzer 1962).

For ordinary (dwarf) stars and for degenerate dwarves t_{md} is typically in the range $10^9 - 10^{10}$ years. The value of t_{md} for neutron stars depends on their complex internal structure, but is probably also very long. Because σ is essentially independent of density in nondegenerate ionized matter, t_{md} is much larger than the age of the universe for very large objects, such as the interstellar medium. Resistive decay of magnetic fields is a slow process for objects of astronomical dimensions.

3.9.3 Dynamos If the fluid is in motion then the first term on the right hand side of (3.9.10) becomes important. In a prescribed steady velocity field this equation is linear in $\vec{B}$, and its solutions will therefore be eigenmodes which either grow or decay exponentially as functions of time. The decay found when $\vec{u} = 0$ is a special case of this behavior. A variety of simple flow patterns may be shown to lead only to decay, but more complex flows may produce exponential growth. This growth is called the dynamo mechanism of magnetic field amplification.

In most real astronomical objects, such as stellar convection zones or the interstellar medium, the flow field is unsteady and turbulent. Such flows are expected to lead to magnetic field amplification by a dynamo process, although it is hard to make a confident prediction for such a complex problem. The characteristic dynamo amplification time t_{da} may be as short as

$$t_{da} \sim \frac{R}{u}, \tag{3.9.13}$$

which will generally be very much less than the age of an astronomical object. If it were proper to speak of eigenmodes of $\vec{B}$ when $\vec{v}$ is a fluctuating quantity, the exponential growth of even a single eigenmode would soon lead to very large $\vec{B}$.

Once the magnetic stress, of magnitude $B^2/8\pi$, becomes comparable to the hydrodynamic stress ρu^2, it is likely that it will alter the flow pattern so that exponential amplification of the magnetic field will cease. This is necessary on energetic grounds, because the fluid motion will then no longer possess enough energy to amplify the field rapidly. It is often assumed that in a turbulent conducting fluid the field will be amplified until this condition is met, so that the field may be estimated

$$\frac{B^2}{8\pi} \sim \rho u^2. \tag{3.9.14}$$

The best observed turbulent conducting flow is probably that of the Solar convective zone. The magnetic field there reaches the values implied by (3.9.14) in sunspots, where it also apparently suppresses the convective motion, in agreement with expectation. Elsewhere on the visible Solar surface the field is much less, in disagreement with (3.9.14). The nondegenerate stars with the largest surface magnetic fields appear to be those whose surfaces are only slightly unstable against convection, or are weakly stable, while an argument based on (3.9.14) would suggest that stars with the most vigorous surface convection should have the strongest fields. In contrast, (3.9.14) appears to approximately describe the magnitude of interstellar fields. The only simple summary is that the real world is not simple, and that this estimate should be used with great skepticism.

3.9.4 Seed Fields Exponential growth by a dynamo process is almost, but not quite, a complete solution to the problem of the origins ot magnetic fields in astrophysics. There must still be a seed field to be amplified, although its magnitude may be extremely small. There are a number of ways in which such a seed field may be produced. It is a general property of pressure gradients in ionized gases that they produce small charge densities and electric fields. To demonstrate this, consider the separate equations of hydrostatic equilibrium for electrons and ions in a uniform gravitational field $\vec{g}$, or subjected to a uniform acceleration $-\vec{g}$:

$$\nabla P_i = n_i m_i \vec{g} + n_i q_i \vec{E} + \vec{F}_i \qquad (3.9.15a)$$

$$\nabla P_e = n_e m_e \vec{g} + n_e q_e \vec{E} + \vec{F}_e. \qquad (3.9.15b)$$

The subscripts i and e refer to the ions and electrons, q, n, and m are respectively the charge, number density, and mass of each species, and $\vec{F}$ is any force other than gravity, inertia, or that of the electrostatic field. The only such force which needs to be considered is the frictional force resulting from collisions between the ions and the electrons. In equilibrium there is no net drift velocity of the two species with respect to each other, so there is no mean frictional force between them, and $\vec{F_i} = \vec{F_e} = 0$.

Addition of (3.9.15a) and (3.9.15b) gives

$$\nabla P = (n_i m_i + n_e m_e)\vec{g} + (n_i q_i + n_e q_e)\vec{E}, \tag{3.9.16}$$

where $P \equiv P_i + P_e$ is the total pressure. If the gas is nearly electrostatically neutral, as will almost always be the case, then $n_i q_i + n_e q_e = 0$ to an excellent approximation. The density $\rho = n_i m_i + n_e m_e$, so that (3.9.16) becomes the ordinary equation of hydrostatic equilibrium (1.3.1):

$$\nabla P = \rho\vec{g}. \tag{3.9.17}$$

Subtraction of (3.9.15b) from (3.9.15a) gives

$$\nabla(P_i - P_e) = (n_i m_i - n_e m_e)\vec{g} + (n_i q_i - n_e q_e)\vec{E}. \tag{3.9.18}$$

Assume a nondegenerate perfect gas equation of state for each species, which in equilibrium are at the same temperature T. If we neglect m_e compared to m_i, the coefficient of $\vec{g}$ is almost exactly ρ. Then the electric field is given by

$$\vec{E} = \frac{\nabla[(n_i - n_e)k_B T] - \rho\vec{g}}{n_i q_i - n_e q_e}. \tag{3.9.19}$$

For the special case of a pure hydrogen plasma $n_i = n_e = n/2$, $q_i = e$, $q_e = -e$, and $\rho = n_i m_i$ (neglecting m_e), so that

$$\vec{E} = -\frac{m_i\vec{g}}{2e}. \tag{3.9.20}$$

The electrostatic potential energy of an ion is then $-\frac{1}{2}$ of its gravitational potential energy. At the surface of the Sun $E \sim 10^{-8}$ Volt/cm, an extremely small field. These fields are found in all gravitating or accelerated plasmas.

The physical origin of this electric field is clear. Nearly all the gravitational force acts on the massive ions, but the electrons contribute substantially to the pressure gradient which opposes it. In order to keep the electrons from being accelerated to infinity by the pressure gradient, a small electric field develops, which confines the electrons, and helps support the ions against gravity. Temperature gradients also produce small thermoelectric fields.

If the matter is moving (nonrelativistically) with respect to an observer, that observer will see a magnetic field

$$\vec{B} = -\frac{\vec{u}}{c} \times \vec{E}. \tag{3.9.21}$$

These fields are very small ($\sim 10^{-27}$ gauss for representative interstellar parameters, and $\sim 10^{-15}$ gauss for the Solar convective zone), but fewer than 50 e-foldings of dynamo amplification will bring them to the observed values.

3.9.5 Field Lines If a fluid is a good conductor ($\sigma \rightarrow \infty$) then (3.9.10) becomes

$$\frac{\partial \vec{B}}{\partial t} = \nabla \times (\vec{u} \times \vec{B}). \tag{3.9.22}$$

The rate of change of magnetic flux Φ through a fixed loop C bounding a surface S may then be found by integration

$$\begin{aligned}\frac{\partial \Phi}{\partial t} &= \frac{\partial}{\partial t} \int_S \vec{B} \cdot \hat{n}\, da \\ &= \int_S [\nabla \times (\vec{u} \times \vec{B})] \cdot \hat{n}\, da \\ &= \int_C (\vec{u} \times \vec{B}) \cdot \vec{dl},\end{aligned} \tag{3.9.23}$$

where $\hat{n}$ represents a unit vector normal to S, da is an element of surface S, $\vec{dl}$ is an element of loop C, and where Stokes' theorem has been used. If the field lines are considered to move with the fluid as if they were frozen into it, then $(\vec{u} \times \vec{B}) \cdot \vec{dl}$ is the rate at which magnetic flux normal to $\vec{dl}$ and to $\vec{u}$ is advected across the element $\vec{dl}$ by the flow. The equality (3.9.23) establishes that the rate of advection of flux across C equals the rate of change of flux through S, so that this picture of magnetic field lines frozen into the fluid is a valid description of the behavior of the magnetic flux in a good conductor.

The concept of frozen magnetic flux is a useful tool. Matter can still bend magnetic field lines (because surface currents flow when it encounters a field, and these currents add to the pre-existing field), even though it does not cross them. If the material stress $P + \rho u^2$ is small compared to the magnetic stress $B^2/8\pi$, even this bending will be slight, and the field will act nearly as an immovable obstacle, confining and guiding the flow of matter.

In the opposite case $P + \rho u^2 \gg B^2/8\pi$, the field lines will be swept along passively with the flow of conducting fluid. This probably describes the behavior of the field in the evolution of stellar interiors. The conservation of flux within a loop implies that if all linear scales contract by a factor $\mathcal{R}$, so that $\rho \propto \mathcal{R}^3$, then the area of a loop varies $\propto \mathcal{R}^{-2}$ and the magnetic field $B \propto \mathcal{R}^2 \propto \rho^{2/3}$. The magnetic stress $B^2/8\pi \propto \rho^{4/3}$, as is expected for a relativistic fluid (**1.9**). This scaling of B may account for the large magnetic fields of neutron stars and degenerate dwarves, although if it is applied to the contraction of stars from interstellar material it predicts impossibly large fields. It is possible that strong fields which violate the condition $P + \rho u^2 \gg B^2/8\pi$ channel the contraction parallel to $\vec{B}$, or prevent contraction until they are reduced by resistive loss or a suitable (anti-dynamo) flow field.

3.10 References

Begelman, M. C., Blandford, R. D., and Rees, M. J. 1984, *Rev. Mod. Phys.* **56**, 255.

Bondi, H. 1952, *Mon. Not. Roy. Ast. Soc.* **112**, 195.

Burger, H. L., and Katz, J. I. 1983, *Ap. J.* **265**, 393.

Clark, D. H., and Caswell, J. L. 1976, *Mon. Not. Roy. Ast. Soc.* **174**, 267.

Eggum, G. E., Coroniti, F. V., and Katz, J. I. 1985, *Ap. J. (Lett.)* **298**, L41.

Ferrari, A., and Pacholczyk, A. G. eds. 1983, *Astrophysical Jets* (Dordrecht: Reidel).

Hoyle, F., and Lyttleton, R. A. 1939, *Proc. Camb. Phil. Soc.* **35**, 405.

Hunt, R. 1971, *Mon. Not. Roy. Ast. Soc.* **154**, 141.

Jackson, J. D. 1975, *Classical Electrodynamics* (New York: Wiley).

Katz, J. I. 1986, *Comments Ap.* **11** in press.

Klein, R. I., Stockman, H. S., and Chevalier, R. A. 1980, *Ap. J.* **237**, 912.

Lada, C. J. 1985, *Ann. Rev. Astron. Ap.* **23**, 267.

Landau, L. D., and Lifshitz, E. M. 1959, *Fluid Mechanics* (Reading, Mass.: Addison-Wesley).

Livio, M. 1986, *Comments Ap.* **11**, 111.

Margon, B. 1984, *Ann. Rev. Astron. Ap.* **22**, 507.

Norman, M. L., Smarr, L., Wilson, J. R., and Smith, M. D. 1981, *Ap. J.* **247**, 52.

Novikov, I. D., and Thorne, K. S. 1973, in *Black Holes* eds. C. DeWitt and B. S. DeWitt (New York: Gordon and Breach), p. 343.

Pringle, J. E. 1981, *Ann. Rev. Astron. Ap.* **19**, 137.

Salpeter, E. E. 1964, *Ap. J.* **140**, 796.

Spitzer, L. 1962, *Physics of Fully Ionized Gases* 2nd ed. (New York: Interscience).

Spitzer, L. 1978, *Physical Processes in the Interstellar Medium* (New York: Wiley).

Zeldovich, Ya. B., and Raizer, Yu. P. 1966, *Physics of Shock Waves and High-Temperature Hydrodynamic Phenomena* (New York: Academic Press).

Chapter 4

High Energy Phenomena

4.1 Accreting Degenerate Dwarves

Degenerate dwarves lie at the border between high energy astrophysics and more conventional astronomy. I have outlined their theory in **1.12**; Shapiro and Teukolsky (1983) give a much more thorough account. There exists a great body of observational data concerning degenerate dwarves, most of which has been obtained with the methods of visible light astronomy. A great deal is known about their spectra and surface chemical composition, magnetic fields, rotation, distribution and motions in space, membership in binary systems, and other conventional astronomical properties.

Disappointingly little is known of the masses and radii of degenerate dwarves. Stellar masses are generally measured from their orbital motions when members of a binary system. For very few degenerate dwarves do data of sufficient completeness and quality exist. The measurement of stellar radii requires either an accurate theory of the emission from stellar atmospheres, or detailed eclipse data. The former is questionable and the latter very rare for degenerate dwarves.

Degenerate dwarf masses and radii are needed for a direct test of the theory of their interiors, but because this theory is grounded on elementary quantum statistics it is probably fair to accept it untested. The absence of these basic data is more serious for the astronomer who is attempting to model quantitatively their observable properties, for whom these are basic parameters.

Some phenomena involving accreting degenerate dwarves may reasonably (if somewhat arbitrarily) be described as belonging to high energy astrophysics. These are the accretion process itself, the problems of their spin and fields, and the dramatic outbursts known as novae. As discussed in **3.5**, accretion from the interstellar medium is generally insignificant, so we are here concerned with degenerate dwarves in binary systems.

4.1.1 Accretion When a degenerate dwarf star is a member of a binary stellar system, and its companion is a nondegenerate star, mass may flow from the companion onto the degenerate dwarf. Córdova and Mason (1983) present a review, and are particularly concerned with the radiation produced by the accretion process itself. The rate of mass flow depends sensitively on the size and evolutionary state of the nondegenerate star and on the separation between the two stars. The most frequently discussed accretion rates are in the range $10^{-10} - 10^{-8} M_\odot$/yr, or $10^{16} - 10^{18}$ gm/sec. Much smaller rates certainly occur when the stars are well separated, but are of little interest; larger rates probably occur but are believed rare. Accreting degenerate dwarves in binary systems were first recognized as variable stars, because accretion flows, whether discs (**3.6**) or more nearly radial (**3.7**), are apparently almost always unsteady, and produce radiation whose intensity varies irregularly on a wide range of time scales. Orbital eclipses and rotation of the accreting star may also lead to periodic variability. In addition, some of these objects are observed to undergo dramatic outbursts known as novae (**4.1.2**); it is likely that they all will eventually (and repetitively) do so.

Accretion onto degenerate dwarves is observationally interesting because it releases a significant amount of energy. The gravitational binding energy of a gram of matter added to the surface is

$$\begin{aligned} \varepsilon c^2 &= \frac{GM}{R} \\ &\approx 2 \times 10^{17} \left(\frac{M}{M_\odot}\right)^2 \text{ erg/gm.} \end{aligned} \tag{4.1.1}$$

The numerical expression is a fair approximation for the mass range $0.4M_\odot \lesssim M \lesssim 1.2M_\odot$, which includes the few measured degenerate dwarf masses. It is generally assumed that most of their masses lie in this range. The problem of mass measurement is particularly treacherous for the most interesting degenerate dwarves, those accreting from binary companions, because emission by the flowing matter confuses the spectroscopic measurement of the orbital Doppler shifts on which mass measurements are based. Measurements of the masses of these stars are generally indirect and controversial. A representative estimate of their mean mass is $M \approx 0.7M_\odot$, so I will henceforth use $\varepsilon c^2 = 10^{17}$ erg/gm, or $\varepsilon \approx 10^{-4}$. Typical accretional luminosities are then in the range $10^{33} - 10^{35}$ erg/sec.

Matter accreted from a binary companion star will often (perhaps almost always) form an accretion disc (**3.6**). Such a disc is expected to form when matter is accreted as a consequence of Roche lobe overflow, and carries with it the large angular momentum of the orbital motion. The cataclysmic variables, a category including novae, dwarf novae, and nova-like variables, are likely examples.

Mass captured from a stellar wind may not form a disc. Significant winds (**1.15**) are produced by luminous stars, whose visible radiation will generally overwhelm that produced by a degenerate dwarf. Degenerate dwarves accreting from the wind of a binary companion may therefore be hard to recognize on the basis of their visible radiation alone, and may only infrequently appear in catalogues of variable stars. The symbiotic stars have been suggested to be examples of this class of objects.

Application of the theory of accretion discs predicts surface effective temperatures (3.6.12)

$$T_e \sim \left(\frac{\dot{M} \varepsilon c^2}{\sigma_{SB} R^2} \right)^{1/4} . \tag{4.1.2}$$

This ranges up to $\sim 10^5$ °K, and the emitted spectrum peaks in the far ultraviolet, with some radiation extending to the soft X-ray ($h\nu \sim 100$ eV) band. Observations in this frequency range are difficult because absorption by photoionization of interstellar matter is large (and depends on the distance of the star and the intervening neutral hydrogen density, both of which are generally poorly known). There may also be significant energy released at the boundary between the disc and the degenerate star, where the temperature may be significantly higher; just how much higher depends on the detailed mechanism of viscous energy release in the boundary layer. Observations show complex variability. The observed energy distributions are in some respects consistent with disc theory, but it alone cannot explain many of the finer details of the data.

If the accreting degenerate star has a significant surface magnetic field (see 4.2.5; $B \gtrsim 10^4$ gauss is probably sufficient), it will disrupt the rotation of an accretion disc near the surface. At least a few, and possibly most, accreting degenerate dwarves have fields this large (a few have much larger fields); fields near the minimum required are hard to detect spectroscopically. Matter deflected or guided by the field falls onto the stellar surface with a velocity

$$v = \sqrt{\frac{\eta G M}{R}}, \tag{4.1.3}$$

where η lies between 1 (the Keplerian orbital velocity) and 2 (the free-fall velocity from infinity, which will be applicable if the matter falls freely along the magnetic field lines from a height $\gg R$). From (4.1.1) v is found to be $\approx 3 \times 10^8$ cm/sec.

Virtually all the kinetic energy of the infalling matter is carried by its ions. Upon entering the stellar atmosphere they are rapidly slowed by Coulomb drag; the proton stopping length is found from (2.2.12) to be $4 \times 10^{20}\ n_p^{-1}\ \text{cm}^{-2}$, where n_p is the density of the protons which slow them. The stopping length is nearly the same for protons and ^{4}He nuclei, which together carry $\approx 97\%$ of the mass and energy. Slowing by electrons is negligible because they are heated so that their thermal velocity far exceeds v, and (2.2.14) is inapplicable.

The stopping length is very short; it is always much less than absorption lengths for photons (outside absorption lines), and is much less than R whenever the rate of accretion is high enough to be of interest. In the limit in which it is zero the slowing of the infalling matter may be regarded as being accomplished by a shock (**3.3**; the thickness of a shock is the thickness required for the particle distribution function to relax to thermal equilibrium). A layer of hot, radiating, shocked matter accumulates between the stellar surface and the accretion flow. Its upper boundary is the accretion shock and its lower boundary is the cooler stellar atmosphere, into which it flows subsonically.

From (3.3.11) and (3.3.17), making the reasonable assumption that the temperature and pressure of the unshocked matter are negligible compared to those of the shocked matter, and taking $\gamma = 5/3$, we obtain the temperature of the matter immediately behind the shock

$$T_s = \frac{3}{16}\frac{\mu m_p v^2}{k_B}, \tag{4.1.4}$$

where μ is the molecular weight ($\mu = 0.6$ for ordinary stellar composition). Then (4.1.3) gives

$$T_s = \frac{3}{16}\frac{\eta\mu m_p}{k_B}\frac{GM}{R} \approx 3 \times 10^8 \left(\frac{M}{M_\odot}\right)^2 \eta \ {}^\circ\mathrm{K}. \quad (4.1.5)$$

Radiation by such hot matter is (usually) largely bremsstrahlung (**2.6.1**); most of the radiated energy is in hard X-rays with energies of tens of KeV, and the shape of the spectrum may (in principle) be used to determine M.

There are a number of complicating effects. If there is a magnetic field with $B \gtrsim 10^7$ gauss, matter at the temperatures indicated by (4.1.5) will rapidly radiate cyclotron harmonic radiation (**2.6.2**) at visible and ultraviolet wavelengths. This process may be important because the radiated intensity at these wavelengths may approach that of a Planck function at the high temperature given by (4.1.5). Elliptically polarized radiation which is probably cyclotron harmonic emission has been observed from a few strongly magnetized accreting degenerate dwarves. Very hot electrons will also be rapidly cooled (2.3.33) by the Compton scattering of lower energy photons emitted by the stellar surface. Both these processes reduce the energy available to be radiated as hard X-rays, and increase the power at much longer wavelengths.

Hard (multi-KeV) X-rays have been observed from numerous accreting degenerate dwarves, including many for which there was no other reason to expect radial infall and shock-heated matter. It is unclear whether magnetic fields sufficient to disrupt disc flow, but too small to observe spectroscopically or polarimetrically, exist in most degenerate dwarves, or whether shocks and high temperatures have another origin. The data are qualitatively consistent with (4.1.5), but it has not been tested in detail.

An interesting question is why accretion flows are usually unsteady. There are two classes of explanations, roughly corresponding to the two classes of turbulence discussed in **1.1**. It is frequently found in hydrodynamics that initially smooth flows are unstable, with small disturbances growing exponentially. Alternatively, accretion may sample a small portion of a much larger and nonuniform flow, and thus be sensitive to local variations in its properties, only whose averages would otherwise be observed. Binary star accretion, whether Roche lobe overflow or from a wind, draws upon the upper atmosphere of the mass-losing star, which is expected to be heterogeneous; the nonuniformity of the Solar atmosphere and corona are known to grow rapidly with height.

4.1.2 Spin and Fields The spin and magnetic field of an accreting object affect and may be determined by accretion. The estimates of **4.2.2** may be applied to degenerate dwarves. Their magnetic fields range from undetectably small values (in a few cases upper bounds are less than 10^4 gauss) to $\sim 10^8$ gauss (Angel 1978). Perhaps 10% of degenerate dwarves are observed to be strongly magnetic, with surface fields $\sim 10^6 - 10^8$ gauss. It is plausible that degenerate dwarves inherit (and compress) the magnetic flux of their nondegenerate progenitors. The origin of this flux is also not understood in detail, but may be related to the presence or absence of surface convection when on the main sequence.

Little is known about the rotation of nonmagnetic degenerate dwarves, because its only readily observable effect is the Doppler broadening of the spectral lines. The lines are also widened by pressure broadening, which increases with surface gravity, so these measurements are not sensitive; lower bounds to rotational periods of tens of minutes exist for a few single stars. In a binary system these measurements would be confused by the spectrum of the companion star.

A magnetic field may make rotation readily observable, by making the radiation spectrum, intensity, or polarization vary across the stellar surface. This is a very large effect when the field guides accretion flow. Single magnetic degenerate dwarves have observed periods ranging from ~ 1 hour to ~ 1 day. In a few cases a period of a century or more has been inferred from the constancy of their radiation, but this argument cannot be conclusive, because if the field is independent of azimuthal angle about the rotation axis (for example, a magnetic dipole aligned with the angular momentum) there will be no rotational modulation of the radiation. Even a period of a day corresponds to a specific angular momentum $\ell \sim 10^{13}$ cm^2/sec, a very low value. The likely explanation is that during the red giant phase of the progenitor star's evolution it rotated roughly as a rigid body; the internal magnetic field may be capable of supplying the necessary rigidity. If the Sun expanded to the dimensions of a large red giant its rotational period would be ~ 1000 years. The much shorter observed periods may be explained if a small fraction of the envelope contracted onto the degenerate core at the conclusion of the red giant phase. Spin periods of a single degenerate star which are $\lesssim 1$ day are unlikely to have undergone significant change since its formation.

Accreting binary magnetic degenerate dwarves are observed to have spin periods ranging from ~ 1 minute to several hours; some of the slower ones rotate in synchronism with the orbital motion. If the magnetic field is not too large then matter accreted from a disc adds its angular momentum to the star; the expected progressive reduction of the spin period has been observed. If the field is very large it may exert a significant electromagnetic torque on the binary companion star, bringing the rotation and orbital motion into synchronism. Such systems are known as AM Hercules stars. It is likely that the spins of accreting degenerate dwarves are determined by these interactions within the binary system.

Very little is known about the spins of non-accreting binary degenerate dwarves, because very few such stars are known (they are hard to detect because the degenerate star is faint and there is no unusual variability to attract attention), and none of these is known to be magnetic. It is possible that their periods may be shorter than those of single degenerate dwarves, even without accretional spin-up, because their nondegenerate progenitors were probably rotating synchronously with the orbital motion, and hence had more angular momentum than single stars.

4.1.3 Novae Historically, the term *nova* was used to describe a star which appeared where none had been known before. In modern usage it refers to a star undergoing a characteristic sudden outburst, in which mass is expelled, but after which the star returns to nearly its pre-outburst state. Novae are distinguished from *supernovae* (**4.3**), in which an entire star explodes, and is completely disrupted or leaves behind only a neutron star or black hole. Gallagher and Starrfield (1978) review the theories and observations of novae.

Nova outbursts are a very heterogeneous class of events. A typical outburst lasts a few months, during which the luminosity is $\sim 10^{38}$ erg/sec, mostly at visible or near-visible wavelengths; the total radiated energy is $\sim 10^{45}$ erg. Spectroscopic studies show that $\sim 10^{-4} - 10^{-5} M_{\odot}$ are expelled, at speeds typically ~ 1000 km/sec. One fact appears to be invariably true, that novae occur in binary stars in which matter flows from an ordinary nondegenerate star (generally a dwarf of mass and luminosity not more than those of the Sun) onto a more compact companion. The spectrum of the companion itself is not directly observed; its faintness, mass, and compact size imply that it is a degenerate dwarf. There is frequently spectroscopic evidence of an accretion disc around the degenerate dwarf, and it may be inferred that the disc is irradiated by ionizing ultraviolet radiation (not directly observable because of interstellar and atmospheric absorption) produced by the degenerate dwarf itself or by the inner regions of its accretion disc.

The outline of a successful model of novae is apparent. The matter accreted by the degenerate star is, like almost all matter in the universe, rich in hydrogen. The burning of hydrogen to helium releases an energy $\varepsilon \approx .005$, so that, integrated over time, thermonuclear reactions produce ~ 50 times as much energy as accretion itself, and are energetically capable of expelling the accreted matter.

The surface layer of hydrogen-rich matter is heated by the conduction of heat from the interior of the degenerate star and by compression as the weight of the overlying matter steadily increases. After a sufficiently massive layer has accreted, its peak temperature will reach a value ($\sim 5 \times 10^6$ °K) at which the release of thermonuclear energy becomes significant, and leads to still further heating.

In contrast to the usual situation in stellar interiors, this release of thermonuclear energy is thermally unstable, meaning that the temperature and the thermonuclear energy generation rate rapidly increase once the energy production becomes significant. The thermal response of an entire star (on time scales slow enough that hydrostatic equilibrium is maintained, an assumption almost always justified) resembles that of a body with negative specific heat, as discussed in **1.5**; this makes its thermal structure stable against perturbations. Thermonuclear energy release in a thin layer is very different. If the layer has mass ΔM, scale height h, and rests atop a stellar core of mass M, radius R, and surface gravity $g = GM/R^2$, and we assume a nondegenerate equation of state and neglect radiation pressure, its internal energy is approximately given by

$$E_{in} = \frac{\Delta M k_B T}{(\gamma - 1)\mu m_p}, \qquad (4.1.6)$$

where μ is the molecular weight, m_p the proton mass, and T the temperature. Its gravitational energy is

$$\begin{aligned} E_{grav} &= -\int \frac{GM\rho(r)}{r}\, 4\pi r^2\, dr \\ &= -\frac{GM\Delta M}{R} + \frac{GM\Delta M h}{R^2} + \cdots, \end{aligned} \tag{4.1.7}$$

where we have taken $\rho(r) = \rho_0 \exp[-(r-R)/h]$ and have assumed the scale height $h \ll R$. Using $h = k_B T/(g\mu m_p)$, we find the total energy of the layer

$$\begin{aligned} E &= E_{in} + E_{grav} \\ &\approx -\frac{GM\Delta M}{R} + \left(\frac{\gamma}{\gamma - 1}\right)\frac{k_B T \Delta M}{\mu m_p}. \end{aligned} \tag{4.1.8}$$

It is evident that the effective specific heat $\partial E/\partial T > 0$, so that significant thermonuclear energy release leads to an accelerating thermal runaway. If the matter is degenerate then h is independent of T to first order. $E_{in} = C_0 + C_1 T + \cdots$, where C_0 and C_1 are constants, and the linear term is contributed by the ions, which remain nondegenerate. The effective specific heat is still positive. Only if the thermonuclear energy production rate is very low (so that it is small compared to the flow of heat from the slowly cooling degenerate core) is the layer stable; in this case, which describes the surfaces of ordinary white dwarf stars, the temperature is effectively set by the stable equations of heat flow.

Once the unstable temperature rise begins it rapidly accelerates because of the steep increase of the thermonuclear energy production rate with T (**1.9.4**). This runaway instability continues until $\partial E/\partial T$ becomes negative, which happens when the assumptions made in deriving (4.1.6) and (4.1.7) no longer apply. Two such assumptions eventually fail, the thin layer approximation $h \ll R$ and the neglect of radiation pressure. As discussed in **1.13**, these both fail when the luminosity L becomes comparable to the Eddington limiting luminosity L_E (1.11.6); the star then settles into a stable state. Its structure resembles that of a luminous giant or supergiant, except that the amount of mass in the envelope is very small ($\lesssim 10^{-4} M_\odot$ instead of $\sim 1 M_\odot$), so the envelope is much smaller. This is the generally accepted outline of the events in a nova outburst, and satisfactorily explains the facts that they only occur in binary systems containing an accreting degenerate dwarf, and that their outburst luminosity $L \approx L_E$.

After the conclusion of the outburst, accretion onto the degenerate dwarf resumes. The accumulation of a fresh hydrogen-rich surface layer may be expected to lead to repeated outbursts, whose intervals depend on the accretion rate and the mass of hydrogen required to initiate them. Typically, these intervals may be in the range $10^4 - 10^6$ years, although a few novae have recurred after intervals of decades. For every nova outburst observed in the last century there are a large number of binary stars which have undergone, and will undergo, such outbursts, but have not done so when astronomers were well equipped to detect them. These binary stars are not luminous (typically $< 10^{-4}$ as bright as the outburst itself), and are much harder to detect. It is likely that each such star erupts $\sim 10^4$ times in its lifetime, a value which may be obtained by comparing the rate of novae in our Galaxy to the number of suitable binary stars. Some of these stars show other kinds of interesting behavior, and are known as dwarf novae, nova-like variables and AM Hercules stars (called "polars" because their strong magnetic fields lead to polarized radiation); together with novae they are called cataclysmic variables.

A number of questions remain. The duration and luminosity of an outburst and the energy available from hydrogen burning ($\varepsilon \approx .005$) require the transformation of only $\sim 10^{-7} M_\odot$ of hydrogen to helium. Calculations typically show that a degenerate dwarf must accrete $\sim 10^{-4} - 10^{-5} M_\odot$ of hydrogen-rich material in order for the thermonuclear runaway to begin, although the numerical value depends on the mass and internal temperature of the accreting degenerate dwarf, and on the accretion rate. It is therefore likely that nearly all the accreted matter is expelled in the outburst; only $\sim 2\%$ need undergo thermonuclear reactions to supply the required energy. Observations of the mass outflow are consistent with this.

It is unclear whether the beginning of a nova outburst should be regarded as an explosion, in which matter is rapidly accelerated to its escape velocity. Such an event would have to take place in the time t_h (1.6.1), which is a few seconds for a degenerate dwarf; if energy is released more slowly the star remains close to hydrostatic equilibrium, and matter is not directly accelerated to escape velocity. It is difficult to produce energy so rapidly by the thermonuclear reactions of hydrogen. The energy required is given by the sum of the gravitational binding energy (4.1.1) and the kinetic energy, and is at least 10^{17} erg/gm. At high temperatures and high energy production rates the burning of hydrogen is catalyzed by heavier elements and proceeds through chains of reactions like

$$^{12}\mathrm{C} + \mathrm{p} \rightarrow {}^{13}\mathrm{N} + \gamma \qquad (4.1.9a)$$

$$^{13}\mathrm{N} \rightarrow {}^{13}\mathrm{C} + e^{+} + \nu_e \qquad (4.1.9b)$$

$$^{13}\mathrm{C} + \mathrm{p} \rightarrow {}^{14}\mathrm{N} + \gamma \qquad (4.1.9c)$$

$$^{14}\mathrm{N} + \mathrm{p} \rightarrow {}^{15}\mathrm{O} + \gamma \qquad (4.1.9d)$$

$$^{15}\mathrm{O} \rightarrow {}^{15}\mathrm{N} + e^{+} + \nu_e \qquad (4.1.9e)$$

$$^{15}\mathrm{N} + \mathrm{p} \rightarrow {}^{12}\mathrm{C} + {}^{4}\mathrm{He}. \qquad (4.1.9f)$$

There are also similar subsidiary reaction chains.

All reaction chains like those of (4.1.9) have the important property that they depend on spontaneous β decays (4.1.9b and e); any sequence of reactions which turns protons into ^{4}He must involve β decays. This sets an upper limit to the rate at which the heavy nuclei can catalyze the reactions. The half lives of ^{13}N and ^{15}O are 598 seconds and 122 seconds respectively; the sum of their mean lives (10^3 seconds) sets a lower limit to the mean time required to complete the cycle (4.1.9). This is clearly much too long to permit the completion of even one cycle in t_h.

At very high temperatures the cycle (4.1.9) is replaced by a series of proton captures without β decays. The most abundant nucleus of those catalyzing (4.1.9) is usually ^{12}C, although after the cycle has been in operation it is largely transformed to ^{14}N. Successive proton captures on ^{12}C produce first ^{13}N, and then ^{14}O (which has a 71 second β-decay half-life); here the captures stop because ^{15}F is not a bound nucleus. Other capture chains lead to similar terminations, although the reaction networks may be more complex. The energy released is generally < 10 MeV per catalytic nucleus. In matter of ordinary stellar composition all elements heavier than helium together comprise no more than 3% of the mass; there are about 10^{21} such nuclei per gram. An energy release of 10 MeV for each catalytic nucleus then amounts to less than 2×10^{16} erg/gm. This is less than the gravitational binding energy (4.1.1), so that a simple thermonuclear explosion does not appear to be possible.

It is often suggested that the abundance of catalytic nuclei in the accreted matter is enhanced over its usual stellar value by mixing with material from the interior of the degenerate dwarf, which is believed to be largely ^{12}C and ^{16}O. When this assumption is made calculations predict satisfactory explosions, and there is some evidence for mixing in the abundances of matter expelled in nova outbursts. It is unclear how this mixing is to be accomplished. The boundary between the heavier elements and the accreted hydrogen is very stable against convective mixing because of the gradient in molecular weight. In addition, direct contact between these layers may be prevented by the presence of a layer of helium left on the surface of the carbon-oxygen core when it was formed, and possibly added to by the products of each nova outburst.

The most conclusive evidence for the explosive nature of nova outbursts would be the observation that they begin with a burst of radiation rising in a time of a few seconds. Such observations do not exist, and would be hard to obtain because nova outbursts are essentially unpredictable; an observer does not know when or where to look. It would be necessary to continually scan much of the sky for transient events. Such observations are new to astronomy, but are possible with electronic detectors which view several steradians with modest angular resolution ($\sim 10^6$ picture elements) and rapid response and signal processing. Spurious terrestrial sources, such as those produced by meteors and airplanes, may be eliminated by comparing the signals from two separated detectors, because they will show a large parallax. The optical study of gamma ray bursts (**4.5**) has similar requirements, and entirely new transient phenomena may be discovered.

It may be possible to explain novae without explosions. As a thermonuclear runaway proceeds on the β-decay time scale ($\sim$ 1 minute) the luminosity increases and the hydrogen-rich layers swell, until $L \approx L_E$. Such luminous stars produce an outflowing wind by a variety of processes (**1.15**), all of which are hard to calculate. In the case of a nova the binary companion provides an additional mechanism. If the swelling luminous star expands to the size of the binary orbit, then the companion will tear through the luminous envelope, violently disrupting it, and probably expelling matter. In some novae mass expulsion continues throughout the visible outburst, so that some continuing mechanism of expulsion is required even if there is an initial explosion. The matter expelled by novae is observed to be very asymmetric, which suggests the influence of the binary companion rather than a process occurring entirely on a spherical star. After the first few minutes the hydrogen burning layer settles down to a steady structure with $L \approx L_E$, whether or not the process began with an explosive ejection of mass.

4.2 Accreting Neutron Stars

Accreting neutron stars are naturally analogous to accreting degenerate dwarves, and display corresponding phenomena. Because the gravitational binding energy of a neutron star is ~ 2000 times greater (**1.12**), the energy released in accretion is greater in proportion, and different physical processes are often important. Accreting neutron stars are therefore much more luminous. Their surface areas are also much smaller, so that one might expect a *qualitative* tendency for their radiation to be at higher frequency; examination of the microscopic physics shows that this is not always so. The visible light produced by an object as small as a neutron star is almost always negligible, so that accreting neutron stars were discovered only by X-ray observation, and most of our information about them comes from these data. Just as for degenerate dwarves, accretion is rapid enough to be of interest only when the neutron star has a close binary companion. Joss and Rappaport (1984) review the extensive literature. The theory of the neutron stars themselves qualitatively resembles that of degenerate dwarves (**1.12**); Shapiro and Teukolsky (1983) discuss it in detail.

Very little is known empirically about the masses and radii of neutron stars, just as for degenerate dwarves. There is one accurate mass measurement, that of the binary pulsar PSR 1913+16, whose mass is determined from relativistic effects on its orbit to be 1.40 $M_{\odot}$. This equals the limiting mass M_{Ch} of a degenerate dwarf (**1.12**). All other mass determinations are very approximate, but all are consistent with this value, and it is plausible (but unproven) that all neutron stars are born with this mass (see **4.3** for a discussion). The theoretical relation between the masses and radii of neutron stars is not accurately known because of uncertainties in the equation of state of matter at high densities; their radii not sensitive to these uncertainties and are often assumed to equal 10 km. This uncertainty in the calculated radii is at least as large as the effects of general relativity, which I will ignore. There are no quantitative measurements of neutron star radii.

4.2.1 Accretion Accreting neutron stars were discovered as luminous X-ray sources. Accretion may occur at a significant rate onto neutron stars with binary companions, just as for degenerate dwarves. The matter may either flow as a consequence of Roche lobe overflow (**3.6**), in which case it carries a great deal of angular momentum, or be captured from a wind, probably with much less angular momentum. In the former case an accretion disc (**3.6**) is formed; in the latter this is less certain, and depends on the details of the flow and the neturon star's magnetic field. Because accreting neutron stars are found from observations of their X-ray emission, their discovery is not impeded by the presence of the luminous companion required to produce a wind, and many examples of such binaries are known; in fact, their high visible luminosity makes it easier to identify them as the visible counterpart of the X-ray source.

The Newtonian gravitational binding energy of a neutron star is

$$\begin{aligned} \varepsilon c^2 &= \frac{GM}{R} \\ &\approx 2 \times 10^{20} \text{ erg/gm.} \end{aligned} \tag{4.2.1}$$

I will henceforth use $\varepsilon \approx 0.2$. If matter flows onto the neutron star through an accretion disc which extends to its surface, the theory of accretion discs applies, and the peak surface effective temperature (3.6.12) is

$$T_e \sim \left(\frac{\dot{M} \varepsilon c^2}{\sigma_{SB} R^2} \right)^{1/4}, \tag{4.2.2}$$

which ranges up to $\sim 10^7$ °K. At this temperature a black body radiates most of its power in X-ray photons of a few KeV energy, qualitatively consistent with the observed X-ray spectra. A boundary layer at the neutron star surface may have somewhat higher temperature. At a given accretion rate T_e is higher than its value for accretion onto degenerate dwarves in proportion to the factor $(\varepsilon/R^2)^{1/4} \propto R^{-3/4} \sim 200$. Because the gravitational binding energies of neutron stars are so large, typical accretional luminosities are in the range $10^{36} - 10^{38}$ erg/sec, much greater than those of accreting degenerate dwarves. The upper end of this range is close to the Eddington limit L_E (1.11.6), and may reflect the trapping of radiation at high accretion rates discussed in **3.7**. The actual flow geometry and radiation transport are probably much more complex.

If a neutron star has a significant surface magnetic field its accretion disc will not extend down to the stellar surface, but will give way to a flow of the accreting matter along the magnetic field lines. This flow will strike the surface with a kinetic energy nearly equal to the gravitational binding energy (4.2.1). Protons will have an energy of about 200 MeV. At this energy Coulomb drag implies a stopping length (2.2.14) of $\approx 8 \times 10^{24}\ n_e^{-1}$ cm^{-2}, corresponding to a column density ≈ 10 gm/cm^2. This exceeds the column density of the entire accretion flow if $L \lesssim L_E$ (**3.7**). There cannot be an accretion shock (unless some other process drastically reduces the stopping length), for its thickness would be greater than that of the flow.

Instead, it is necessary to consider the deceleration of the infalling ions in the surface layers of the neutron star. They heat the stellar atmosphere, which then radiates. Because the stopping length is long, much of the accretion power is deposited deep in the atmosphere, at optical depths $\gtrsim 1$. The radiation field in these deep layers approaches that of a black body, and emerges from the surface with an effective temperature approximately given by (4.2.2). The dilute and optically thin upper layers of the atmosphere are also heated by

Coulomb drag, and reach higher temperatures because they radiate less effectively. Their principal cooling mechanism may be Compton scattering (**2.3**) of the radiation from the deeper layers, but they also contribute more energetic (harder) X-rays to the spectrum by bremsstrahlung (**2.6.1**). The result, calculated by Alme and Wilson (1973), is a complex multi-component spectrum, which is at least qualitatively consistent with observations of X-ray sources.

It is remarkable that accretion onto neutron stars is expected to produce lower energy X-rays than accretion onto degenerate dwarves, because the neutron star T_e (4.2.2) is much less than the degenerate dwarf accretion shock T_s (4.1.5). This expectation is borne out observationally, if the hard X-rays attributed to the degenerate dwarf accretion shocks are compared to the radiation from accreting neutron stars. Of course, radiation from degenerate dwarf surfaces at the temperature (4.1.2) is characterized by a much lower temperature.

The physics of accreting neutron stars is complicated by the presence of magnetic fields. These channel the flow onto a small portion of the neutron star's surface, so that the optical depth may be high even if $\dot{M} \ll \dot{M}_E$ (3.7.7). If $B \gtrsim 10^{12}$ gauss the magnetic field contributes a large cyclotron opacity (**2.6.2**). If $\hbar\omega_B \gtrsim k_B T$ (2.6.16) then the field also affects the microscopic physics of the matter, changing electron wave functions from nearly free particle states (whose gyration about the field may be described classically) to states strongly confined in directions perpendicular to the field. The result is that the matrix elements for all electronic processes differ from their usual values. At the highest fields $\hbar\omega_B \gg k_B T$, and the electrons are nearly all in states in which their magnetic quantum number (their angular momentum parallel to the field) $m = 0$; their spins are also aligned. Calculations of both the microscopic physics and the overall hydrodynamic flow and radiation transport around magnetic neutron stars are complex and difficult, and the results depend on the values of several parameters.

4.2.2 Spin and Fields It was observed that the intensities of many (but not all) luminous X-ray sources are regularly pulsed. The first pulse periods discovered were a few seconds, but they have now been observed to span the range $10^{-1} - 10^{3}$ seconds. These periods are very stable, but they show regular Doppler shifts attributable to binary orbital motion. Optical observations of these X-ray sources show ordinary nondegenerate stars with the same orbital periods, but with oppositely phased Doppler shifts. These stars must be binary companions of the X-ray sources, and are sources of mass for accretion. Pulsed X-ray sources with these properties are sometimes called "binary X-ray pulsars." This name is unfortunate because they are very different from ordinary pulsars (**4.4**), a few of which are also binary and a few of which are X-ray sources.

The outline of a theory of these objects (which I will call accreting magnetic neutron stars or pulsating X-ray sources) was immediately apparent. The accretion flow onto a magnetic neutron star is channelled by its field, so that the power released per unit area and radiated intensity vary across its surface. As it rotates an observer sees varying aspects of the surface, and therefore an intensity and spectrum modulated with the spin period. The minimum required value of the surface magnetic field is probably $\sim 10^{8}$ gauss. Observations of pulsars suggest that most neutron star magnetic fields are $\sim 10^{12} - 10^{13}$ gauss; features in the spectrum of a few accreting magnetic neutron stars have been identified with cyclotron radiation lines produced in fields of this magnitude. An accreting neutron star with a magnetic field much less than 10^{8} gauss would not be expected to show modulated X-ray emission. Many unmodulated X-ray sources exist, and may be explained as neutron stars with small magnetic fields, or possibly as black holes.

The chief theoretical difficulty is the description of the interaction of the accreting matter with a neutron star's magnetic field. This problem is hard because it depends on the accretion flow, which is not known in detail (whether an accretion disc or a more nearly radial flow, the actual distribution of density and velocity are not quantitatively calculable), and also on complex magnetohydrodynamics and plasma physics. If these difficulties are slighted it is possible to make estimates which are consistent with much of the data.

Assume that there is an abrupt transition from an inner to an outer flow regime at a radius r_A. For $r < r_A$ the magnetic stress $B^2/8\pi$ exceeds the fluid stress (3.1.6) $\rho u^2 + P \approx \rho u^2$. Because hot ionized gases are very good electrical conductors, they cannot readily flow across magnetic field lines (**3.9**). If the magnetic stress much exceeds the fluid stress, the magnetic field configuration will be nearly that of its vacuum state, and the fluid will only be able to flow parallel to the vacuum field lines. It will follow those field lines which extend from the star to r_A, where the fluid can bend the field. This region in which the magnetic stress is dominant is called the magnetosphere. If $r_A \gg R$, where R is the stellar radius, the accretion flow will be guided onto a small area at the magnetic poles.

For $r > r_A$ the fluid stress exceeds the magnetic stress, and it is assumed that the fluid flows as if there were no field. Because the fluid is a good conductor its inward motion sweeps in the stellar field, which is largely excluded from the fluid for $r > r_A$. If the infall is spherically symmetric the stellar field is completely excluded from this region. We will assume that here the matter flow is that of an accretion disc, as described in **3.6**.

Because we cannot calculate the complex processes occurring at $r \approx r_A$, we assume that there the accreting matter is captured from the accretion disc, along with its Keplerian orbital angular momentum at that radius. Its mass and angular momentum are added to those of the neutron star. The radius r_A is found by comparing the magnetic and fluid stresses

$$\frac{B^2}{8\pi} = \rho u^2. \tag{4.2.3}$$

For B substitute the equatorial magnetic field of a dipole moment μ, $B = \mu/r_A^3$, and for u the Keplerian orbital velocity $u = \sqrt{GM/r_A}$. The disc density is less certain, because it depends on the viscous stress or radial component of velocity, which are unknown even to order of magnitude. We write $\rho = \varsigma\dot{M}/(4\pi r_A^2 v)$, where $\varsigma = 1/\sqrt{2}$ would correspond to spherically symmetric free radial infall and $\varsigma \gg 1$ to slow radial motion through a disc. For $r < r_A$ the flow may approximate radial infall, with $\varsigma \sim 1$. Then r_A is found from (4.2.3):

$$r_A = \left(\frac{\mu^2}{2\varsigma\dot{M}\sqrt{GM}}\right)^{2/7}. \tag{4.2.4}$$

For representative neutron star parameters ($\mu = 10^{30}$ gauss cm^3, $\dot{M} = 10^{17}$ gm/sec, $M = 1.4M_\odot$, and $\varsigma = 1$), $r_A \approx 3 \times 10^8$ cm. For a magnetic degenerate dwarf ($\mu = 10^{33}$ gauss cm^3, $\dot{M} = 10^{17}$ gm/sec, $M = 0.7M_\odot$, and $\varsigma = 1$), $r_A \approx 1.6 \times 10^{10}$ cm. In both cases $r_A \gg R$. For a degenerate dwarf in a compact binary system r_A may approach the distance to the companion star, in which case it may prevent the formation of any accretion disc; this is believed to describe the AM Hercules stars (**4.1.2**).

The minimum equatorial surface dipole field required in order to channel the accretion flow ($r_A > R$) may be obtained from (4.2.4), and is

$$B = \frac{(2\varsigma\dot{M})^{1/2}(GM)^{1/4}}{R^{5/4}}. \tag{4.2.5}$$

For $\varsigma = 1$, $\dot{M} = 10^{17}$ gm/sec, and typical neutron star parameters this is $\sim 5 \times 10^7$ gauss; for degenerate dwarf parameters it is $\sim 10^4$ gauss.

The specific angular momentum accreted is

$$\begin{aligned} \ell &= \sqrt{GMr_A} \\ &= (GM)^{3/7}\mu^{2/7}(2\varsigma\dot{M})^{-1/7}. \end{aligned} \tag{4.2.6}$$

The rate of change of the accreting star's angular momentum is

$$\begin{aligned}\frac{d(I\Omega)}{dt} &= \dot{M}\ell \\ &= (GM)^{3/7}\mu^{2/7}\dot{M}^{6/7}(2\varsigma)^{-1/7},\end{aligned} \tag{4.2.7}$$

where I is the moment of inertia and Ω the spin frequency.

If $\Omega t_h \ll 1$ (1.6.1) then the fractional variation in Ω is much more rapid than that in I. Taking I to be constant, expressing the result in terms of the period $P \equiv 2\pi/\Omega$, and assuming that the accreted angular momentum is parallel to that already possessed by the star, gives

$$\dot{P} = -\frac{P^2(GM)^{3/7}\mu^{2/7}\dot{M}^{6/7}}{2\pi(2\varsigma)^{1/7}I}. \tag{4.2.8}$$

This relation may be directly compared to the observed P and $\dot{P}$, if estimates are made for I (from models of neutron star interior structure), ς, μ, and M, and if $\dot{M}$ is calculated from the observed luminosity. The data are approximately consistent with this theory (Joss and Rappaport 1984).

The significance of this consistency should not be exaggerated. Because r_A and ℓ are only weak functions of the various parameters, their dependence is not tested by the approximate agreement between the data and the theory. The magnetic moment μ is presumably different for different neutron stars. Without independent measures of it, which generally do not exist, the predicted close correlation of $\dot{P}$ with $P^2\dot{M}^{6/7}$ would not be found even if (4.2.8) held exactly. Essentially the same results would be obtained if the theory of the magnetospheric boundary were replaced with the simple assumption of a constant ℓ, guessed at or fitted to the data, replacing (4.2.8) by the simpler expression

$$\dot{P} = -\frac{P^2\dot{M}\ell}{2\pi I} \tag{4.2.8$'$}$$

It is possible to distinguish accreting degenerate dwarves from accreting neutron stars on the basis of their values of $\dot{P}$, because both (4.2.8) and (4.2.8′) predict that at a given luminosity $\dot{P}$ is about 10^3 times smaller for degenerate dwarves. One important empirical conclusion is also established, that the accreted matter does carry significant angular momentum, although not necessarily enough to form a disc at $r \gg r_A$.

The actual behavior of P is more complicated than (4.2.8) would indicate. The previous discussion assumed that the magnetosphere exerts no torque on the flow at $r > r_A$. This must be correct for $\Omega = 0$, but is unlikely to be so if Ω is large. It is plausible that the critical angular frequency Ω_c at which this torque equals the accretion torque is that for which the Keplerian angular velocity at r_A equals the velocity of rotation of the accreting star's magnetic field lines there, which may be thought of as rigid spokes extending from the star:

$$\Omega_c = \sqrt{\frac{GM}{r_A^3}}. \tag{4.2.9}$$

Substituting r_A from (4.2.4) gives

$$\Omega_c = (GM)^{5/7} \mu^{-6/7} (2\varsigma \dot{M})^{3/7}. \tag{4.2.10}$$

When the spin frequency of the accreting star reaches this value it is likely that there is no further mean torque, and that its frequency and period stabilize. If ς and $\dot{M}$ vary, as is likely, Ω_c will fluctuate about the mean $\langle \Omega_c \rangle$, and Ω will decrease when instantaneously $\Omega_c < \Omega$.

For the neutron star parameters we have used (and $\varsigma = 1$) Ω_c corresponds to $P \sim 2$ sec, while for degenerate dwarf parameters it corresponds to $P \sim 20$ minutes. A number of accreting magnetic neutron stars (including, but not only, those with Ω comparable to the expected Ω_c) show a very variable $\dot{P}$ which even changes sign, implying that sometimes angular momentum flows from the star to the disc. In those cases for which $\Omega \sim \Omega_c$ this is consistent with theory.

Very few useful data exist for accreting magnetic degenerate dwarves because their larger I implies a small $\dot{P}$. The shortest observed periods of $\sim$ 1 minute require (if the theory is correct) $\mu \lesssim 3 \times 10^{31}$ gauss cm^3, instead of the $\mu \sim 10^{33}$ gauss cm^3 of strongly magnetized degenerate dwarves; this is consistent with the absence of evidence for strong fields in their spectrum and polarization, but still permits the field to exceed the minimum (4.2.5).

Very slowly rotating accreting neutron stars spin up very rapidly. In a few cases the characteristic e-folding time of the period $t_{su} \equiv P/|\dot{P}|$ is less than a century. Ordinarily it is hard to observe such rapid processes because at any given time the number of objects passing through such a brief stage in their evolution is small. Equivalently, it is possible to argue that if the duration of this stage is 100 years, for every one we see there must be an additional 10^8 objects in the Galaxy which passed through this stage in its age. These numbers exceed by a large factor the number of suitable binary systems containing neutron stars.

The likely resolution of this problem is that slowly rotating accreting neutron stars are made from faster ones, so that each neutron star passes through the slowly rotating (but rapidly accelerating) phase many times. There are several possible mechanisms for reducing Ω. Nearly all these accreting magnetic neutron stars accrete from a luminous companion with a stellar wind. If the wind is suitably heterogeneous, the angular momentum may even reverse direction, changing the sign of (4.2.8), and reducing Ω while accretion continues. This may explain the fact that some neutron stars with small Ω occasionally increase their periods instead of reducing them, even though $\Omega \ll \Omega_c$.

The mean trend of the observed X-ray sources, even the slowly rotating ones, is for P to decrease, so that long spin periods are not maintained by alternate episodes of accretion of matter with opposite directions of angular momentum. Slow rotation may be restored if the neutron stars lose angular momentum when they are not observed; that is, when they are not accreting. If Ω much exceeds Ω_c the spin may prevent accretion (rather than just reversing the sign of the torque) because this implies that the centrifugal force on matter on the field lines at r_A exceeds the attraction of gravity. The only way to obtain $\Omega > \Omega_c$ for a long spin period is for μ to be very large or for ρ to be very small. It is conceivable that these stars have extraordinarily large μ, but it is not possible for them to change μ between their episodes of spin-up (when their rates of spin-up suggest that μ is comparable to that for other accreting neutron stars) and spin-down. Spin-down may occur if the external supply of matter and ρ are reduced, which is described by small $\dot{M}$ in (4.2.10), although accretion does not take place and no X-rays are emitted.

The rate of spin-down during these dark episodes is even harder to calculate than the rate of spin-up when accretion takes place, and there are no data. It may be comparable in magnitude to (4.2.8), with $\dot{M}$ now some characteristic mass flow rate above the magnetosphere, but this will be very slow if $\dot{M}$ is small, as is implied by the requirement that Ω_c (4.2.10) decrease below the already low Ω. In this case the number of dark systems is at least $\sim (\Omega_{c0}/\Omega)^2 \sim 10^5$ times the number of bright ones spinning up, where Ω_{c0} is the value of Ω_c determined by the accretion rate during the bright spin-up phase. This number is daunting, if not impossible.

It is also of interest to consider the rate of spin-down of a rotating magnetic dipole. In vacuum the power radiated is

$$P = \frac{2\ddot{\mu}^2}{3c^3}, \qquad (4.2.11)$$

analogously to (2.6.1), where μ refers separately to each component of the vector $\vec{\mu}$. This is negligible for neutron stars rotating as slowly as most of the accreting ones. However, in a medium which can propagate waves with speed v_w, to which the oscillating dipole moment couples, an analogous derivation shows that the radiated power is

$$P \sim \frac{\ddot{\mu}^2}{v_w^3}, \tag{4.2.12}$$

provided that the interaction with the waves remains linear. If v_w describes magnetohydrodynamic waves in the accretion disc or other surrounding fluid it may be orders of magnitude less than c, and very rapid spin-down is conceivable. The applicability of (4.2.12) and the quantitative magnetohydrodynamics of this flow are speculative.

4.2.3 X-Ray Bursts The matter accreted by a neutron star will generally be of ordinary stellar composition, and capable of releasing a fraction $\varepsilon \approx .007$ of its rest mass in thermonuclear energy. In contrast to the case of degenerate dwarves, where thermonuclear burning of accreted matter releases more energy than accretion itself ($\varepsilon \approx 10^{-4}$), for neutron stars thermonuclear energy might be thought insignificant compared to accretion ($\varepsilon \approx 0.2$). Although the total energy released is relatively small, it is still of great interest because it produces characteristic X-ray bursts which are analogous to nova outbursts (**4.1.3**).

X-ray bursts typically last a few seconds to a minute, have a peak luminosity $L \approx 10^{38}$ erg/sec, and recur on time scales of an hour to days. Their emitted spectrum is roughly that of a black body at the temperature implied by the surface area of a neutron star and the observed luminosity.

Thermonuclear energy release of matter accreted onto a neutron star may be thermally unstable for the same reasons discussed in **4.1.3** for degenerate dwarves. Such thermonuclear flashes are the natural explanation of X-ray bursts. The brief durations of the bursts, shorter than the half-lives of the β-unstable catalysts in the CNO cycle (4.1.9), imply that hydrogen burning is not their chief source of energy. Helium burning by the reaction (1.9.30) may proceed essentially arbitrarily rapidly, and is the likely source of energy. Quantitative calculations (Joss and Rappaport 1984) have shown that this model accounts fairly well for the observed properties of X-ray bursts.

It is striking that the accreting magnetic neutron stars discussed in **4.2.2** do not show X-ray bursts, while the X-ray sources which do have bursts do not show regular rotational periods. The probable explanation, which is supported by detailed calculations, is that when the accreting matter is concentrated onto the magnetic poles its temperature is higher and its thermonuclear fuel is consumed steadily as it is accreted, rather than accumulating until an outburst occurs. The accreting neutron stars which have bursts then have lower magnetic fields, probably below the value (4.2.5).

Most of the X-ray sources in globular clusters show X-ray bursts, which establishes that they are accreting neutron stars with small magnetic fields. Earlier suggestions that they were black holes cannot account for this phenomenon. The other bursting X-ray sources are concentrated in the central bulge of our Galaxy, evidence that they are part of the same population of old ($\sim 10^{10}$ years) stars as the globular clusters. They are almost certainly members of binary stellar systems, for there is no other adequate source of matter for accretion. Their companion stars are usually not observed in visible light, and are therefore probably faint old stars of low mass. In contrast, the companions of the accreting magnetic neutron stars are usually luminous young ($\sim 10^7$ years) massive stars, although there are a few exceptions. This may be interpreted to mean that the magnetic field of a neutron star decays in a time between 10^7 and 10^{10} years.

4.2.4 γ-Rays All the neutron star phenomena we have so far discussed are thermal, in the sense that the particle distribution functions are Maxwellian in the frame of the mean fluid velocity, and no acceleration of energetic particles is required. The only exception to the Maxwellian distribution functions was the presence of fast ions, representing the penetration of accreting fluid into a stellar surface. Their acceleration is only that of gravity, and their slowing is satisfactorily explained as the Coulomb drag of test particles within a thermal plasma.

The X-ray source Cygnus X-3 is known to produce γ-rays with energies approaching 10^{16} eV (Samorski and Stamm 1983). This enigmatic object appears to be a binary star with a 4.8 hour orbital period, but displays neither a spin period nor X-ray bursts, and therefore need not contain an accreting neutron star. Various authors have suggested that it contains a rapidly spinning pulsar (**4.4**). Because pulsars are a nonthermal phenomenon, the presence of very energetic γ-rays, and the acceleration of the $\sim 10^{16}$ eV particles required to produce them, is not astonishing (Eichler and Vestrand 1984).

Soon afterward, γ-rays of similar energy were reported from the accreting magnetic neutron stars Vela X-1, Hercules X-1, and LMC X-4 (Protheroe, *et al.* 1984, Baltrusaitis, *et al.* 1985, Protheroe and Clay, 1985), and γ-rays of $\sim 10^{12}$ eV from Hercules X-1 and 4U0115+63 (Dowthwaite, *et al.* 1984, Chadwick, *et al.* 1985). The observed signals were not far above the threshold of statistical significance, but appear to be genuine. This *was* astonishing, because there had been no reason to suspect the existence of any nonthermal phenomena in any accreting magnetic neutron star. The absence of nonthermal phenomena was generally attributed to the presence of dense accreting thermal plasma, which was assumed to prevent the development of the high electrostatic potentials required to accelerate energetic particles and to explain pulsar activity.

The four neutron stars from which γ-rays were observed have spin periods (1.24, 3.61, 13.5, and 283 seconds) which span the range from fast to slow. They do not differ strikingly from other objects of this class, except that they all have smaller values of $|\dot{P}|$ than other objects of corresponding P and $\dot{M}$ (Joss and Rappaport 1984, Chadwick, *et al.* 1985). This might imply smaller magnetic fields, but in two cases this is contradicted by the values of the cyclotron frequency inferred from the X-ray spectrum. A significant fraction of the total power of these systems appears to take the form of energetic particle acceleration. If these objects do not systematically differ from other accreting magnetic neutron stars then more sensitive observations will discover many more high energy γ-ray sources.

Given a charged particle of 10^{16} eV energy, it is easy to produce γ-rays of comparable (but somewhat smaller) energy. An electron of this energy will transfer nearly all of its energy to a photon by Compton scattering (**2.6.3**) if the photon were originally of microwave or higher frequency. It is difficult to accelerate such energetic electrons in strong magnetic fields because they radiate both synchrotron radiation and curvature radiation, the process analogous to synchrotron radiation but in which the electron is accelerated as it follows a curved magnetic field line. A proton of high energy will produce γ-rays (and neutrinos) by colliding with protons at rest:

$$\begin{aligned} \mathrm{p} + \mathrm{p} &\rightarrow \text{pions} + \text{other hadrons} \\ \pi^o &\rightarrow \gamma + \gamma \\ \pi^+ &\rightarrow \mu^+ + \nu_\mu \\ \mu^+ &\rightarrow \mathrm{e}^+ + \bar{\nu}_\mu + \nu_e \\ \pi^- &\rightarrow \mu^- + \bar{\nu}_\mu \\ \mu^- &\rightarrow \mathrm{e}^- + \nu_\mu + \bar{\nu}_e. \end{aligned} \tag{4.2.13}$$

The electrons and positrons may subsequently produce additional γ-rays by Compton scattering.

It is hard to explain the acceleration of 10^{16} eV charged particles, even protons. At least three classes of mechanisms have been considered. It might be that accreting matter avoids some of the neutron star's open magnetic field lines, and that on these field lines it could accelerate energetic particles like a pulsar (though apparently without producing radio pulses), while the thermal accretion process proceeds on other field lines. Unfortunately, pulsar theory predicts an upper limit to the accelerating potential (4.4.20) which falls far short of that required, a shortfall which is worse in proportion to P^2 for long spin periods. Further, pulsar activity is driven by neutron star rotational energy, and this model would imply a rate of spin-down orders of magnitude faster than the observed rate of changes in spin period (which also usually have the wrong sign).

A second mechanism (Chanmugam and Brecher 1985) applies pulsar theory to an accretion disc, as was suggested by Blandford (1976) and Lovelace (1976) in order to explain energetic particle acceleration in quasars. This mechanism is uncertain because the magnetic field magnitude and distribution in a disc are quite unknown. If it is effective, it should take place in all accretion discs, including those around nonmagnetic neutron stars, black holes, and degenerate dwarves. Pulsar particle acceleration depends on the scarcity of free charged particles which could "short out" the accelerating field; it is unclear how this extraordinarily good vacuum could be obtained near an accretion disc in a mass-transfer binary, where ionized gas flows in abundance.

A third mechanism involves acceleration within the magnetosphere. Particles on open field lines will escape immediately, but protons gyrating around closed field lines may bounce back and forth between the magnetic poles many times. If a suitable density of plasma waves (for which there is yet no evidence) is excited by the infalling thermal plasma, energetic particles may gain energy by scattering from the waves (**2.5.2**), thus indirectly tapping the energy of accretion. In order to produce a large power in the accelerated particles it is necessary that they draw upon the gravitational acceleration of matter into the deep potential close to the neutron star.

The ingenuity of theoretical astrophysicists will surely develop more models.

4.3 Supernovae

4.3.1 Phenomenology Supernovae have been observed with the naked eye since ancient times, although the records of prescientific observers do not always make it possible to distinguish supernovae from unusually bright novae or even from the more spectacular comets. Trimble (1982, 1983) presents an extensive review.

Supernovae are catastrophic explosions which destroy an entire star. The debris of these explosions typically have a mass of 1 − 10 $M_\odot$, and are expelled with velocities in the range $10^8 - 2\times10^9$ cm/sec. Supernovae are extraordinarily luminous, often as bright in visible radiation as an entire galaxy, and are therefore readily detected on photographs of distant galaxies. This intense radiation lasts for several months, fading gradually. The total energy radiated is generally in the range $10^{49} - 10^{51}$ erg. Some nearby supernovae have been seen in the daytime. Supernovae in our own Galaxy are rare, with the last four generally accepted events observed in the years 1006, 1054, 1572, and 1604, all before the invention of the telescope. Supernovae were originally termed *novae*, because they appeared as new stars where none had been known before. The distinction between novae (**4.1.3**) and supernovae followed the observation in 1885 of a supernova in the Andromeda galaxy, and the later recognition of the great distances of external galaxies, which implied the enormous luminosity of supernovae.

Studies of distant galaxies have collected data on hundreds of supernovae, some of which have been bright enough to permit spectroscopic observations of good resolution. Because these data are of high quality, a nearby supernova, though spectacular, might not be particularly informative. The mean rate of occurrence of supernovae is estimated (uncertainly and controversially) as one per ~ 30 years in a galaxy like our own. Most of the supernovae in our Galaxy are not observed because of the great extinction (scattering and absorption) of visible light by interstellar dust. Between us and a similar point on the far side of the Galaxy this extinction may reduce the intensity by a factor $\sim 10^{-16}$. This would make even a supernova unobservable in visible light, though not necessarily so in the infrared, where the extinction coefficient is much less (roughly in proportion to the photon frequency).

Supernovae appear to fall into at least two distinct classes, with a few unusual events fitting neither class. By noting the kinds of galaxies in which they occur, and the populations of stars making up these galaxies, it has been shown that Type I supernovae occur in populations of very old stars, while Type II supernovae occur in populations containing much younger stars. This has important implications for their origins.

The expanding debris of a supernova are naturally called its remnant (**3.4**). These remnants are generally observed as nonthermal radio sources, believed to result from the synchrotron radiation of relativistic electrons accelerated by the interactions of the expanding remnant with the interstellar medium (**2.5**). Within this cloud there may be a compact remnant—a neutron star or a black hole. In a few cases a neutron star is known to be present because it is observed as a pulsar (**4.4**; the most famous is the pulsar within the Crab nebula, the remnant of the supernova of 1054). Most other supernovae, including those observed in historic times, show no direct evidence for a neutron star. It is possible, or even likely, that these supernovae completely disrupted their progenitor star, and left no compact remnant. An isolated black hole or neutron star which was not a pulsar would be hard to detect, and could be present.

4.3.2 Visible Radiation The observation that supernovae are luminous for several months has important implications. If an exploding star suddenly expands from a radius r_0, mean density ρ_0, and internal energy density $\mathcal{E}_0$ at a characteristic velocity v, and its matter is characterized by an adiabatic exponent γ (1.9.7), then these quantities scale with time approximately as

$$r \sim vt \tag{4.3.1a}$$

$$\rho \sim \rho_0 \left(\frac{vt}{r_0}\right)^{-3} \tag{4.3.1b}$$

$$\mathcal{E} \sim \mathcal{E}_0 \left(\frac{vt}{r_0}\right)^{-3\gamma} . \tag{4.3.1c}$$

The total internal energy of the expanding fluid, ignoring radiative losses, is

$$\begin{aligned} E &\sim \mathcal{E} r^3 \\ &\sim E_0 \left(\frac{vt}{r_0}\right)^{-3(\gamma-1)} . \end{aligned} \tag{4.3.2}$$

The internal energy is gradually transformed to kinetic energy, as a pressure gradient accelerates the matter. The exponent γ is generally between 4/3 and 5/3, so that when $vt \gg r_0$ the remaining internal energy $E \ll E_0$.

The radiated energy must be drawn from the internal energy of hot matter. If there were no continuing source of internal energy after the initial explosive event the energy available for radiation would decrease rapidly for $t \gtrsim r_0/v$. Because copious radiation is observed for $t \sim 10^7$ seconds, and v is observed spectroscopically to be in the range $10^8 - 2 \times 10^9$ cm/sec, this argument implies $r_0 \sim 10^{15} - 10^{16}$ cm. This inferred radius is extraordinarily large for a star (the largest red giants are smaller than 10^{14} cm), and it does not appear to be possible to suddenly heat and explode such a large configuration, even if it were to exist. The only explanation of the observed long duration of luminous supernova radiation is continuing replenishment of the internal energy of the expanding matter.

There are at least two ways in which internal energy may be replenished. If the exploding star is surrounded by an extended gas cloud, the expanding debris will collide with the cloud, and a portion of its kinetic energy will be thermalized as internal energy. This process is analogous to the blast wave formed which the supernova debris later collide with the surrounding interstellar medium (**3.4**), but the distance and time scales are very much less. Because the density and optical depth of the cloud are high, the radiant energy must diffuse through the cloud, and the emergent spectrum may resemble that of a black body, although the temperature of the matter immediately after it passes through the shock will be very high. Equating the power released in the shock to that radiated by a black body leads (for plausible parameters) to a surface effective temperature in the range 5,000 – 10,000 °K, consistent with observations of supernova spectra. This model requires that the supernova be surrounded by a gas cloud of radius $10^{15} - 10^{16}$ cm, which must be attributed to mass loss by the pre-supernova star, probably in the few hundred years immediately prior to the supernova.

An alternative model for the replenishment of internal energy suggests that the supernova debris contain a large quantity of radioactive material whose decay supplies the required energy. The processes

$$\begin{aligned} {}^{56}\mathrm{Ni} + e^- &\rightarrow {}^{56}\mathrm{Co} + \nu_e \\ {}^{56}\mathrm{Co} + e^- &\rightarrow {}^{56}\mathrm{Fe} + \nu_e \end{aligned} \tag{4.3.3}$$

proceed with half lives of 6.1 and 79 days, respectively (including a small contribution to the rate of the latter process by positron emission), with the electrons captured from bound atomic levels. These decay rates, measured under terrestrial conditions, are essentially the same whenever $T \lesssim 3 \times 10^6$ °K, so that the K-shell electrons remain bound to the unstable nuclei. In the early stages of a supernova explosion interior temperatures will be higher than this value, but simple consideration of the energy contained in the radiation field shows that by the time $r \gtrsim 10^{13}$ cm ($t \gtrsim$ 1 hour − 1 day) T will drop sufficiently that the terrestrial decay rates are applicable. This hypothesis is attractive because it offers a natural explanation of the nearly exponential decay of intensity observed for many supernova light curves (those of Type I), with decay rates close to those of the nuclear decays (4.3.3). It has the additional advantage of explaining the relatively high cosmic abundance of ^{56}Fe, which must be copiously produced by supernovae in which (4.3.3) is important, and is implied by the expected efficient nucleosynthesis of ^{56}Ni under conditions occurring in some models of supernova interiors. The decay by spontaneous fission of ^{254}Cf (half life 61 days) has also been suggested, but would imply the production of excessively large quantites of rare actinide and fission product isotopes.

It is apparent that an exploding star which did not produce a large quantity of ^{56}Ni, and was not surrounded by a suitable gas cloud, would not produce the long duration of visible light emission characteristic of observed supernovae. If we had not known of the empirical properties of supernovae, we probably would not have predicted either of these processes, and would not expect supernovae to be the long lived optical objects they are. It is possible that there are stellar explosions which do not produce the familiar visible supernovae. These may be expected to show much briefer bursts of radiation, whose duration ($t \sim r_0/v$) ranges from a few hours (if $r \sim 10^{13}$ cm, characteristic of a red giant) to as short as a second (if

$r \lesssim 10^9$ cm, characteristic of a degenerate dwarf). Because of their small size, the peak power these explosions radiate in visible light is also very much less than that of observed supernovae, although large compared to that of ordinary stars. Such events would be essentially undetectable in conventional photographic searches for extragalactic supernovae. Most of their radiation would be at ultraviolet or X-ray wavelengths, which are much harder to observe.

At present, the only way to infer the existence of stellar explosions which are not visible supernovae would be to find a supernova remnant or neutron star young enough and close enough that the event which formed it should have been observed in historic times as a visible object, but was not. There is no clear evidence for such objects. One young supernova remnant, known as Cassiopeia A and estimated to be 300 years old, was not observed in visible light, but it is a matter of controversy what limits may be set on its visible luminosity, in part because of significant interstellar extinction in its direction. Finding relatively faint and brief stellar explosions is an important challenge for instruments designed to study transient astronomical events.

4.3.3 Mechanisms A mechanism for supernovae must both trigger a sudden catastrophic event and produce the observed explosion. These requirements are distinct, and should be considered separately. Because at least two distinct types of supernovae are observed, occurring in different populations of stars, at least two mechanisms are required.

The most obvious mechanism of supernova explosion is the release of thermonuclear energy. Most stars (the exceptions are neutron stars, and any degenerate dwarves which have, contrary to expectation, iron interiors) would explode if all their thermonuclear energy could be released suddenly. Under ordinary conditions this cannot happen, because thermonuclear energy generation is stably self-regulating. As discussed in **4.1.3**, this is not the case for thin shells of thermonuclear fuel, but these do not contain nearly enough energy to power a supernova. In a degenerate stellar interior, such as that of a degenerate dwarf, giant, supergiant, or other very evolved star, the pressure is nearly independent of temperature: $P = P_0[1 + \mathcal{O}(k_B T/\epsilon_F)^2]$, where ϵ_F is the electron Fermi energy. Then the hydrostatic structure and gravitational binding energy E_{grav} are similarly nearly independent of temperature. The internal energy of electron-degenerate matter is of the form

$$E_{in} = C_0 + C_1 T + \cdots, \tag{4.3.4}$$

with $C_1 > 0$. The C_1 term is contributed by the ions, which remain nondegenerate. The total energy is of the form

$$\begin{aligned} E &= E_{in} + E_{grav} \\ &= C + C_1 T + \cdots . \end{aligned} \tag{4.3.5}$$

The effective specific heat $\partial E/\partial T$ is positive, in contrast to the usual case of nondegenerate stellar interiors (**1.5**). Thermonuclear energy generation is then unstable, and leads to an accelerating thermal runaway. This runaway ends only when the temperature rises enough that the electrons become nondegenerate, at which high temperature the reaction rates may be so rapid that the energy release in the time t_h (1.6.1) required for the star to expand may be sufficient to disrupt it.

This thermonuclear explosion mechanism has been suggested to occur with helium, carbon, oxygen, and perhaps heavier fuels (mixtures of elements near silicon). In each case the details of the conditions under which it begins depend on the prior evolution of the star, and are not readily summarized. Whether the star is actually disrupted depends on its density and on the details of its structure and the properties of its matter. If the thermonuclear runaway is insufficiently violent the star will settle into a state of steady energy release, rather than being disrupted, and will remain there until the burning fuel is exhausted. At the high densities ($\sim 10^9 - 10^{10}$ gm/cm^3) characteristic of these runaways, hot stellar matter rapidly radiates neutrinos, whose luminosity far exceeds that radiated as photons. The neutrino emission determines the course of stellar evolution, and may perhaps drain enough energy from a thermonuclear runaway to prevent stellar disruption.

A quite different mechanism of triggering a supernova depends on the properties of the equilibrium equation of state of matter at high density or high temperature. As discussed in **1.5**, a star with mean adiabatic exponent $\gamma < 4/3$ is unstable, and will collapse or explode on the time scale t_h. If γ decreases below 4/3 as the result of a decrease in pressure in a star previously in equilibrium, it can only collapse. If $\gamma \approx 4/3$ throughout a star, then a significant decrease of γ in its core may reduce the mean γ below 4/3. There are several ways in which this may possibly take place. In a very massive and luminous nondegenerate star the core is always too massive ($M_{core} > M_{Ch}$; **1.12**) to be supported by degeneracy pressure, and its temperature will steadily rise throughout its evolution. When it reaches $\sim 5 \times 10^8$ °K the thermal equilibrium density of electron-positron pairs becomes significant. The creation of these pairs is endothermic because of their rest mass, and reduces the particle kinetic energy and radiation energy which contribute to the pressure, thus reducing both P and γ. Endothermic nuclear processes have similar effects in dense degenerate matter at yet higher temperatures, as stable heavy nuclei (like ^{56}Fe) disintegrate to yield ^{4}He and neutrons, which are less tightly bound. This may occur in the degenerate cores of stars massive enough for nuclear reactions to have produced ^{56}Fe, but less massive than those in which pair instability occurs. In cold degenerate matter at even higher densities the electron Fermi energy and chemical potential are high enough that it becomes energetically favorable for the electrons near the Fermi surface to be captured on ordinarily stable nuclei. By reducing the equilibrium electron density at a given mass density, this reduces P and γ. This process may occur in any degenerate star (or degenerate core) whose mass is sufficiently close to M_{Ch}.

All of these equation of state instabilities produce collapse, rather than explosion, but collapse may lead to explosion. The rapidly increasing density and temperature of hydrodynamic collapse lead to rapidly increasing thermonuclear reaction rates, and possibly to explosion. The collapse of degenerate matter may instead continue until it reaches neutron star density. Then two mechanisms exist to reverse the collapse and possibly to produce an explosion. It is known (**1.12**) that the limiting mass of cold neutron stars $M_{Ch}^{ns} > M_{Ch}$, because nucleons strongly repel each other so that the equation of state of matter at neutron star densities is very "stiff." This stiff equation of state may lead to a hydrodynamic bounce, sending a shock into the outer layers of the star, and perhaps expelling them. Alternatively, a bounce may be produced when neutrinos emitted by the hottest and densest inner regions of the collapse are absorbed or scattered further out, depositing their energy and momentum in matter whose gravitational binding is less tight. An outgoing shock may also lead to the explosive release of thermonuclear energy in shock-heated matter.

The problem of supernova mechanisms is complex. Each of these mechanisms has been the subject of detailed calculations, and many of controversy. The problem is complicated by couplings among the various mechanisms, because collapse may trigger an explosion, and the possibly explosive release of thermonuclear energy may instead lead to collapse.

Most of these mechanisms (the exceptions are pair instability and perhaps the thermal disintegration of stable nuclei) require the accumulation of a degenerate core whose mass essentially equals $M_{Ch} = 1.40 M_\odot$. If this core collapses without expelling or accreting a significant amount of mass (which would require the expulsion of the entire nondegenerate envelope), a neutron star will be formed with mass M_{Ch}. All measured neutron star masses are consistent with this hypothesis.

The fact that supernovae are observed in a population of old stars is itself interesting. In this population the minimum mass (M_{Ch}) required for collapse or explosion poses a problem, for all stars with $M > 0.9 M_\odot$ terminated their evolution in the distant past. Only less massive stars have long enough lives on the main sequence to be living, and dying, today. Mass transfer in binary stars is the probable resolution of this problem. A degenerate dwarf may perhaps accrete enough matter from its companion, presently swelling as its core evolves, to bring the degenerate mass up to M_{Ch}. To do this it must avoid expelling the accreted material in nova outbursts (**4.1.3**), but this may be possible at high accretion rates, when a steady luminous envelope may be formed instead of intermittent outbursts.

It should be remembered that these three phenomena are distinct: a visible supernova, the violent end of a star, and the formation of a compact object (neutron star or black hole). The relations among them are unclear. The supernova of 1054, which produced a neutron star and the Crab nebula, was all three. Many other supernovae appear not to have produced compact objects, and it is possible that compact objects are sometimes produced without visible supernovae. Stars may perhaps explode without producing either a supernova or a compact object. It is certain only that stars must die to produce either of the other two phenomena.

4.4 Pulsars

4.4.1 Phenomenology Pulsars were an unexpected discovery (Hewish, *et al.* 1968) by a radio telescope designed to study rapid variations in the intensities of radio sources resulting from the refraction of radio waves in the turbulent interplanetary medium (interplanetary scintillation). At its low operating frequency (81.5 MHz) large arrays of dipole antennas make a cheap and sensitive telescope. Pulsars are relatively intense at these frequencies. Most importantly, while conventional radio observations integrate the power received from sources presumed to be steady in order to increase the ratio of signal to noise, this instrument was designed and operated to detect rapid variations of intensity. It is nearly a general rule that the most interesting discoveries are made by the first instrument to perform a new kind of observations, entering a new range of wavelengths, or becoming sensitive to other properties of a signal (spectrum, temporal variation, polarization, angular resolution, or statistics) which were not previously observed.

Pulsars received their name because their radiation is pulsed with a very regular period. Most of the periods observed are ~ 1 second, but they span the range 0.0015 – 4 seconds. Pulsars are searched for and found as sources of pulsed radio radiation, particularly at low frequencies ($\lesssim$ 400 MHz), but a few of those with shorter periods ($\lesssim$ 0.1 second) also emit pulses at visible, X-ray, or γ-ray frequencies. Detailed models are very uncertain, but all of these radiations require the motion of very energetic electrons (or positrons) in strong magnetic fields.

Gold (1968) argued that the magnitude and stability of pulsar periods could best be explained if they were rotating magnetized neutron stars. The spin period of such a massive body would be little affected by outside influences. Neutron stars are believed to have crystalline outer layers, so that they should rotate as rigid bodies.

Unless rapidly accreting, torques on them would be very small, making their spin periods stable. The values of pulsar periods are too short to be the spin or vibrational periods of degenerate dwarves, but too long to be the vibrational periods of neutron stars. If magnetized, with a magnetic field which is not azimuthally symmetric about the spin axis, (for example, the field of a dipole moment not aligned with the angular momentum vector), then any radiation process associated with the magnetic field should vary with the azimuthal angle like a searchlight, and an observer will see a radiation intensity pulsed with the rotational period. Such a rotating magnetized star in vacuum will emit magnetic dipole radiation at its spin frequency and will lose rotational energy, so that its spin period will slowly but steadily increase. This prediction was soon confirmed, and this model accepted.

The regularity of a pulsar period P is usually described by a quality factor Q:

$$Q \equiv \frac{1}{\dot{P}}. \tag{4.4.1}$$

Measured values of Q are in the range $10^{12} - 10^{19}$, with typical values $\sim 10^{14} - 10^{15}$. Pulsar rotation rates are therefore among the most stable clocks in the universe, rivalled or exceeded only by the excitation of metastable nuclear or hyperfine states. It is possible to measure such large values of Q with clocks of lesser quality factors because the Q of a clock is defined differently. Unlike a pulsar, the mean frequency of a clock does not generally drift with time. Considered as an oscillator, a clock's $Q \equiv \omega/\Gamma$, where ω is its angular frequency and Γ its damping rate. Q may be relatively small, but for an atomic clock ω may generally be regarded as absolutely constant. If continuously and stably excited, a clock permits the measurement of a mean frequency to an accuracy much greater than Γ. In contrast, a pulsar's spin is continuously and monotonically slowing, so that in the course of an extended period T of observation its spin frequency

Ω changes by a relatively large amount $\Delta\Omega/\Omega \sim T/(PQ) \gg \Gamma/\omega$. Practical values of T/P are $\sim 10^8$ for typical pulsars, and $\sim 10^{11}$ for those few with millisecond periods (those with the highest Q),

The instantaneous power radiated by each component of an oscillating magnetic dipole moment $\vec{\mu}$ in vacuum is

$$\frac{dE}{dt} = \frac{2|\ddot{\vec{\mu}}|^2}{3c^3}. \tag{4.4.2}$$

For a dipole moment rotating at angular frequency Ω, with angle α between its spin axis and $\vec{\mu}$, the total time-averaged power is

$$\left\langle \frac{dE}{dt} \right\rangle = \frac{2\mu^2\Omega^4 \sin^2 \alpha}{3c^3}, \tag{4.4.3}$$

where μ is the magnitude of the dipole moment. For a neutron star of radius R, the magnetic moment and the magnitude B_0 of the field at the magnetic equator are related by

$$\mu = B_0 R^3. \tag{4.4.4}$$

The radiated power is drawn from the kinetic energy of rotation.

It is an excellent approximation for all but the very fastest pulsar (and probably accurate to $\sim 10\%$ even for it) to consider the rotational flattening to be small, and to treat the moment-of-inertia tensor as a constant scalar I, which is estimated to be $\sim 10^{45}$ gm cm^2. Then the rate of spin-down is given by

$$-I\Omega\dot{\Omega} = \left\langle \frac{dE}{dt} \right\rangle, \tag{4.4.5}$$

or

$$\dot{\Omega} = -\frac{2B_0^2 R^6 \Omega^3 \sin^2 \alpha}{3c^3 I}. \tag{4.4.6}$$

From measured values of Ω and $\dot{\Omega}$, using theoretical values of R and I for a neutron star, and taking $\sin\alpha \sim 1$, B_0 may be estimated. For most pulsars it is $\sim 10^{12}$ gauss, but for a few (chiefly those with very short spin periods) it is much lower, in one case $\lesssim 10^8$ gauss. There are strong observational selection effects, because a rapidly spinning pulsar with a large field will soon slow down, and is therefore unlikely to be observed. In the case of the Crab pulsar, surrounded by a luminous supernova remnant filled with energetic electrons radiating synchrotron radiation at frequencies from radio to X-ray, the power ($10^{38} - 10^{39}$ erg/sec) inferred from (4.4.5) is sufficient to supply the electrons if the spin energy is efficiently converted to particle acceleration.

If $\dot{\Omega}$ is known, then it is possible to define a characteristic slowing or spin-down time

$$t_{sd} \equiv \frac{\Omega}{|\dot{\Omega}|}. \tag{4.4.7}$$

The result (4.4.6) may be integrated to express the actual age t_a of a pulsar in terms of t_{sd}, Ω, and Ω_0, its spin frequency at birth. However, it is better first to generalize (4.4.6) to the form

$$\dot{\Omega} \propto -\Omega^n, \tag{4.4.8}$$

where n is called the braking index. If n is constant the age is

$$t_a = \frac{t_{sd}}{n-1}\left[1 - \left(\frac{\Omega}{\Omega_0}\right)^{n-1}\right]. \tag{4.4.9}$$

It is apparent from (4.4.9) that the relation between the observed t_{sd} and the actual age t_a depends on the pulsar's unknown initial spin Ω_0. There are two limiting cases. If $\Omega \ll \Omega_0$ then $t_a = t_{sd}/(n-1)$, while if the measured Ω is very close to Ω_0 then $t_a \ll t_{sd}$. It is likely that pulsar ages are widely distributed between these limits.

The age of a pulsar may be measured directly only if other information is available. In a few cases a pulsar is clearly associated with a supernova remnant, and must have the same age. The best example of this is the Crab pulsar, born in the supernova of 1054. Most pulsars are believed to be old enough ($\gg$ 30,000 years) that any supernova remnant is now undetectable, and it is possible that some pulsars are born quietly, without either a visible supernova or a remnant of expelled gas. It is also possible to make statistical estimates of the mean age of pulsars based on their spatial distribution within the Galaxy, their places of formation, and their velocities. The resulting ages are $\sim 3 \times 10^6$ years.

It is not known with what spins pulsars are born. If they were to inherit the specific angular momentum of their progenitor stars they would spin very rapidly, with periods $\lesssim 0.001$ sec. This angular momentum is large enough to interfere with collapse to neutron star densities, so that the rate of collapse would be limited by the rate at which angular momentum is removed from the contracting stellar core (in analogy with the formation of stars from rotating interstellar clouds **3.2**). However, neutron stars are expected to form from the degenerate cores of giant stars. If approximately rigid rotation is maintained within such a star (**4.1.2**) the specific angular momentum of the core would be extremely small, and even when collapsed to neutron star densities the spin period would be very much less than that of pulsars. For pulsars the truth must lie between these extremes, although some neutron stars may be born with very long spin periods.

Some pulsars have empirical values of t_{sd} much longer than the ages estimated from their distribution within the Galaxy. These pulsars must have been born with nearly their observed periods (usually ~ 1 second), and will cease emitting (for reasons not well understood) without significant change in their periods. In the case of the Crab pulsar $t_{sd} \approx 2500$ years, so that $t_a \sim t_{sd}$. Its initial period is unlikely to have been very much shorter than its present value of 0.033 second, because there is no evidence for the large kinetic energy lost by a very rapidly rotating neutron star, which would be expected to have observable effects on the supernova remnant. When its period has lengthened to a more typical value ~ 1 second, $\Omega \ll \Omega_0$ will be satisfied. An unknown fraction of pulsars presently observed with typical periods were similarly born with much shorter ones.

Application of (4.4.9) to the Crab pulsar shows that unless n was significantly larger in the past than it is now, its initial period was about half its present period, consistent with the energy argument. We can conclude that pulsar periods at birth span at least the range $\sim 0.01 - 1$ second, but how they are distributed within this range is uncertain.

A few pulsars have very short (1 – 10 milliseconds) spin periods and low magnetic fields. It has been suggested that their rapid spins resulted from accretion from a disc (**4.2.2**) long after their formation as more slowly rotating neutron stars. This is unproven, but it may be that no pulsars are born with periods shorter than ~ 0.01 second. The existence of very short periods at birth would imply the availability of a great deal of neutron star rotational energy, and the possibility of efficient production of gravitational waves in the aspherical collapse which produced the rapidly rotating neutron star.

It is much more difficult to measure $\ddot{\Omega}$ than $\dot{\Omega}$. Some pulsars have irregular variations in $\dot{\Omega}$, attributable either to fluctuations in the relaxation of their interiors to the slowly declining Ω, or to fluctuations in their actual spin-down torque. A few pulsars also show small impulsive *increases* in Ω, called "glitches," which briefly interrupt the steady spin-down. These are attributed to sudden relaxation of their structure, perhaps as a result of fracture of their solid crust or of a sudden increase in its coupling to their superfluid interior. For two pulsars $\ddot{\Omega}$ is large enough to be measured reliably. The measurement is best expressed in terms of the braking index n, which may be written:

$$n = \frac{\ddot{\Omega}\Omega}{\dot{\Omega}^2}. \tag{4.4.10}$$

The observed values of n are $2.515 \pm .005$ for the Crab pulsar, and $2.83 \pm .03$ for PSR1509-58 (Manchester, *et al.* 1985).

A value $n = 3$ might be expected from (4.4.6), but the smaller observed values may be explained if $\sin\alpha$ is increasing on roughly the same time scale as t_{sd}. Because $\sin\alpha$ cannot increase beyond 1, this explanation would imply that as $\Omega/\Omega_0 \to 0$, $n \to 3$. The present significant deviation of n from 3 then suggests that for the Crab pulsar Ω_0 did not greatly exceed Ω. This is consistent with the energetic argument and with (4.4.9) evaluated for the empirical n, which implies a spin period at birth of 0.019 second. Alternatively, the vacuum magnetic dipole theory may be rejected as a quantitative description of spindown. There are good theoretical reasons for doing so (**4.4.2**), in which case the deviation of n from 3 may be attributed to gradual changes in the electrodynamic structure of the pulsar's magnetosphere.

It has occasionally been suggested that rotating neutron stars might emit gravitational quadrupole radiation (for which $n = 5$). They can do so only if their mass distribution is not azimuthally symmetric about their rotation axis. There is no reason to expect such an asymmetry, because a rotating fluid body in equilibrium will be azimuthally symmetric, and the mechanical strength of a neutron star's solid outer layers can support only very small asymmetries against its gravity. There is no evidence of any gravitational radiation produced by the spin of any pulsar. The orbital motion of a pulsar (or any other mass) in a binary system is expected to produce gravitational radiation, which has been observed and the theory verified for the binary system containing the pulsar PSR1913+16. The pulsar makes this observation possible by providing an extraordinarily accurate orbiting clock.

Very few pulsars (about 1% of those known) are observed to have binary companions. In contrast, perhaps half of all stars are binary. Two hypotheses may explain this fact, and it is likely both contribute. It is observed from their slow motion across the sky that pulsars have large space velocities (the dispersion of each component is $\approx$ 120 km/sec; Lyne, *et al.* 1982), yet their distribution in space is that of massive young stars which have much smaller velocities. It is inferred that the event which formed the pulsar endowed it with a recoil velocity. The observed velocities would have been sufficient to expel the pulsar from a binary system, unless the orbital radius was fairly small. Yet if a pulsar has a moderately close binary companion it is likely to accrete. The accreted matter becomes ionized and is a good electrical conductor, and would probably prevent the development of the large electric fields (**4.4.2**) required to accelerate energetic particles and produce pulsar emission. These two hypotheses of recoil and plasma suppression of pulsar emission satisfactorily (though qualitatively) explain the data. The observed binary pulsars have companions which are either of low mass and luminosity or (PSR1913+16) are believed to be another neutron star. In either case the rate of mass flow should be negligible.

The radio pulses emitted by pulsars have a very high brightness temperature T_b, defined from the Planck function (1.7.13) as the temperature of a black body source which produces the observed flux density F_ν:

$$k_B T_b = \frac{F_\nu c^2}{2\nu^2}, \tag{4.4.11}$$

where $h\nu \ll k_B T_b$ has been assumed. Using the pulse width Δt to estimate the area of the emitting region $A \sim (c\Delta t)^2$ leads to numerical estimates of T_b as high as $\sim 10^{26}$ °K, or $k_B T_b \sim 10^{22}$ eV.

No black body can have such a temperature. Emission by individual energetic particles of energy E can generally produce $k_B T_b \sim E$, so this is also not a feasible explanation of the large measured T_b. Only collective or coherent processes are satisfactory. The production of high brightness temperatures by the collective motion of large numbers of electrons is a familiar phenomenon. For example, a dipole antenna radiating 1 KW into a 1 KHz bandwidth at a frequency of 1 MHz has $k_B T_b \sim 10^{18}$ eV. In pulsars coherent emission is probably the result of the bunching of radiating electrons by a plasma instability.

Because pulsars, neutron stars, and supernova remnants are (at least sometimes) born in the same events, it is of interest to compare the birth rates of each. Of the several historic supernovae, only the one of 1054 has an associated pulsar, that in the Crab nebula. It has been suggested that other supernova remnants contain pulsars which are not observed because their beam of radiation is never directed toward us, and that therefore our counts of pulsars underestimate their number by a factor of $\sim 5 - 10$. This is unlikely because the Crab supernova remnant has a characteristic appearance, filled with electrons accelerated by the pulsar and emitting synchrotron radiation. Most other supernova remnants do not appear to have this central source of very energetic electrons. Rather than most pulsars being undetectable because of beaming, it is more likely that most supernova remnants do not contain pulsars, and most supernovae do not produce them. It is also possible that some pulsars are born without a supernova or without a remnant of expanding gases.

4.4.2 Electrodynamics The model of pulsars as rotating magnetized neutron stars satisfactorily explains much of their phenomenology, but needs further development to begin to explain their radiation. Acceleration of relativistic particles is required to account for the observed radiation at all frequencies, as well as the supply of particles to the surrounding region so vividly observed in the Crab nebula. These particles must also be bunched to explain the radio frequency radiation, but this is not required for that at higher frequencies, where the brightness temperatures are much less.

The electrodynamics of a rotating magnetized neutron star is a very difficult problem, as yet incompletely solved. Goldreich and Julian (1969) developed its basis, and many authors have contributed to its further and more controversial elaboration. The problem is very complicated, and no solution is complete or free of difficulty. See Michel (1982) for a review of the literature and numerous hypotheses.

Assume that the magnetic dipole moment is parallel (or anti-parallel) to the angular momentum vector $\vec{\Omega}$. Such an aligned moment cannot lead to pulsed radiation, but is a more tractable model of the complex problem of a misaligned moment. The neutron star is also assumed (with justification) to be a good electrical conductor, so that in the frame of its rotating matter the electric field $\vec{E}' = 0$ (3.9.7). Expressing this in terms of the fields in an inertial frame (3.9.8)

$$\vec{E} + \left(\frac{\vec{\Omega} \times \vec{r}}{c} \right) \times \vec{B} = 0, \tag{4.4.12}$$

where $(\vec{\Omega} \times \vec{r})/c$ is the velocity vector of the rotating matter. Then $\vec{E} = -\nabla\Phi$ determines the electrostatic potential Φ within the neutron star. If the neutron star were surrounded by a perfect vacuum Φ could be calculated in the vacuum region.

Any surface currents are likely to be dissipated by resistivity in the outermost layers of the neutron star, so that $\vec{B}$ just inside the star is essentially the same as the assumed dipole field outside, which in spherical coordinates is

$$\vec{B} = \frac{B_0 R^3}{r^3}(2\hat{r}\cos\theta + \hat{\theta}\sin\theta). \tag{4.4.13}$$

Then (4.4.12) yields at $r = R$:

$$\vec{E} = -\frac{B_0 R \Omega \sin\theta}{c}(-\hat{r}\sin\theta + 2\hat{\theta}\cos\theta), \tag{4.4.14}$$

where $\vec{\Omega}$ has been taken to be directed along the $\theta = 0$ axis. The tangential component of $\vec{E}$ must be continuous across the neutron star surface. This is sufficient to determine Φ (up to a constant) by integrating in θ:

$$\Phi(r=R) = \frac{B_0 R^2 \Omega}{c}(\sin^2\theta + C). \tag{4.4.15}$$

In the vacuum outside the star Φ is determined by Laplace's equation $\nabla^2\Phi = 0$, with (4.4.15) as a boundary condition. The angular dependence is a Legendre polynomial, which must be of degree 2 to match (4.4.15). Then

$$\Phi(r,\theta) = \frac{B_0 R^5 \Omega}{r^3 c}\left(\sin^2\theta - \frac{2}{3}\right) + C'. \tag{4.4.16}$$

The electric field is

$$\vec{E}(r,\theta) = \frac{B_0 R^5 \Omega}{r^4 c}\left[3\hat{r}\left(\sin^2\theta - \frac{2}{3}\right) - 2\hat{\theta}\sin\theta\cos\theta\right], \tag{4.4.17}$$

and

$$\vec{E}\cdot\vec{B} = -\frac{4B_0^2 R^8 \Omega}{r^7 c}\cos^3\theta. \tag{4.4.18}$$

Within the neutron star $\vec{E}\cdot\vec{B} = 0$ (4.4.12). In order that the component of $\vec{E}$ normal to the surface be discontinuous (as is implied by the discontinuity in $\vec{E}\cdot\vec{B}$), there must be a surface charge. Within this surface charge layer $\vec{E}\cdot\vec{B}$ will vary continuously, and the charge will be freely accelerated by the component of $\vec{E}$ parallel to $\vec{B}$. The result (4.4.18) shows that (except at the magnetic equator) this component is not zero, and in fact is comparable in magnitude to $|\vec{E}|$. For the usual pulsar parameters ($B_0 \sim 10^{12}$ gauss, $\Omega \gtrsim 1$ sec^{-1}) the electrostatic force on a charged particle will be many orders of magnitude greater than that of gravity, so that any free charges present will be accelerated and will fill a magnetosphere with plasma. This plasma will be a source of charge and current, invalidating the vacuum results (4.4.16) – (4.4.18).

The net charge density required to reduce $\vec{E} \cdot \vec{B}$ to zero is given by

$$\begin{aligned}\rho_e &= \frac{\nabla \cdot \vec{E}}{4\pi} \\ &= -\frac{\vec{B} \cdot \vec{\Omega}}{2\pi c} \\ &= \frac{B_0 R^3 \Omega}{2\pi r^3 c}(\sin^2\theta - 2\cos^2\theta),\end{aligned} \tag{4.4.19}$$

where $\vec{E}$ is taken from (4.4.12), $\vec{B}$ from (4.4.13), and $\nabla \times \vec{B} = 0$ because $\vec{B}$ is the field of a static dipole (expected to be a good approximation for $r \ll c/\Omega$). If the magnetosphere is filled with this net charge density it is electrodynamically equivalent to the stellar interior. The electric field (4.4.12) is just that induced by (and sufficient to maintain) corotation of the plasma at the spin frequency of the neutron star. If (4.4.12) and (4.4.19) were to hold everywhere around a pulsar it would not accelerate particles or emit radiation.

The particle density implied by (4.4.19) is very small, corresponding near the surface of a typical pulsar to $\sim 10^{11}$ cm^{-3}. This is $\sim 10^9$ times less than the particle density near the neutron star surface in a typical accretion flow. If there is any accretion, even at a rate much too low to produce observable luminosity, a very small polarization of the accreting plasma is sufficient to satisfy the condition $\vec{E} \cdot \vec{B} = 0$, to "short out" the electric field, and to prevent particle acceleration. Such objects cannot radiate as pulsars. This polarization is necessary to justify the description of plasma flow by the equations of neutral fluid magnetohydrodynamics (**3.9**), and is present in most astrophysical flows.

The magnetic field lines may be divided into two classes. The closed field lines do not extend further than a distance c/Ω, called the corotation radius (or the radius of the speed of light cylinder), from the neutron star, and return to the opposite magnetic hemisphere. Open field lines originate near the magnetic poles and extend to distances $\gg c/\Omega$, corresponding to the wave zone of a rotating dipole in vacuum, and bend back only at extremely large distances where they encounter the fields and stress of the supernova remnant or the interstellar medium. Only field lines crossing the neutron star surface within an angle θ_0 of the magnetic pole will be open, where $\theta_0 \approx \sqrt{\Omega R/c}$. Because charged particles are trapped on magnetic field lines, and are therefore not easily lost from closed field lines, the closed field lines are believed to readily fill with the net charge density (4.4.19), and to form a corotating magnetosphere of little interest.

Charges accelerated from the neutron star surface on the open field lines may freely stream out to the supernova remnant. It is often assumed that everywhere (except for a narrow accelerating region) on these field lines $\vec{E} \cdot \vec{B} = 0$, so that they are equipotentials. The net charge density (4.4.19) might suggest that particles of only one sign of charge are present on the open field lines, because they all begin near the poles where $\sin\theta \ll \cos\theta$. Although ρ_e has a single sign in these polar regions, particles of both signs of charge must be lost in equal numbers in order that the pulsar not accumulate a large net charge (the loss of a very few particles, if all of the same sign of charge, would produce a monopole potential sufficient to stop further loss).

If $B_0 > 0$ $(\vec{B} \cdot \vec{\Omega} > 0)$ the surface potential (4.4.15) has minima at $\theta = 0$ and at $\theta = \pi$. Electrons are therefore accelerated along field lines near the axis. Positively charged particles are accelerated along field lines which leave the star at θ close to θ_0 and $\pi - \theta_0$, where the potential achieves its highest open field line value. If $B_0 < 0$ $(\vec{B} \cdot \vec{\Omega} < 0)$ the positively charged particles flow along the axis, and the negatively charged particles near θ_0 and $\pi - \theta_0$. The current loop is closed by currents in the interior of the neutron star and in the supernova remnant. The total potential available for acceleration is

$$\begin{aligned}\Delta\Phi &\approx \frac{B_0 R^2 \Omega \sin^2\theta_0}{c} \\ &\approx \frac{B_0 R^3 \Omega^2}{c^2} \\ &\approx 1.3 \times 10^{13} \frac{B_{12}}{P^2} \text{ Volts},\end{aligned} \tag{4.4.20}$$

where $B_{12} \equiv B_0/(10^{12}$ gauss) and P is the spin period in seconds. This gives the maximum energy to which a particle may be accelerated. Depending on how this potential drop is distributed along the field line, and on radiative losses (which may be large for electrons and positrons), the actual particle energy may be much less.

If an abundant supply of charged particles of both signs were available for acceleration the accelerating potential would be reduced by the plasma space charge (4.4.19), while if none were available the accelerating potential would have its maximum vacuum value. Pulsar magnetospheres must be between these limiting cases.

The magnetized surfaces of neutron stars are believed to be strongly cohesive, because the magnetic field confines the electronic wave functions (in the two dimensions normal to the field), and thus increases their Coulomb binding to the nuclei (which are expected to be Fe in a neutron star surface). The matter is estimated to reach zero pressure at a density (which depends on the magnetic field) $\rho \sim 10^4$ gm/cm^3, and therefore to resemble a solid metal rather than a plasma. The energies required to detach electrons and ions are quantitatively uncertain, but probably of the order of kilovolts (much greater than surface thermal energies), and substantially larger for nuclei than for electrons. The supply of charged particles from such a surface is very uncertain, and may be difficult. It is worth noting that Φ (4.4.16), $\vec{E}$ (4.4.17), and ρ_e (4.4.19) depend on the sign of B_0 (the sign of $\vec{B} \cdot \vec{\Omega}$). Because the difficulties of removing ions and electrons from the surface differ, it may be that neutron stars with different signs of $\vec{B} \cdot \vec{\Omega}$ have different properties. Conceivably, one sign leads to a pulsar and the other to some other object, perhaps a γ-ray source.

If no particles are available to be accelerated there can be no radiation. Sturrock (1971) and Ruderman and Sutherland (1975) have suggested a solution to this problem. A γ-ray of energy $E \gg 2m_e c^2$ can create an electron-positron pair in a strong magnetic field. Without a field this process could not simultaneously conserve energy and momentum, but a strong field can take up momentum and permit one photon pair creation (similarly, one photon annihilation may occur, and is the dominant pair annihilation process for $B > 10^{13}$ gauss). In the intense electric field (4.4.17) both electrons and positrons are rapidly accelerated along magnetic field lines. Because the field lines are curved the particles are also accelerated in a direction normal to their velocity and radiate many photons (the radiation resulting from their acceleration parallel to $\vec{B}$ is negligible). At sufficiently high energy this curvature radiation may itself produce pairs. The result is a multiplication of the γ-ray and charged particle density, which continues until the accelerating potential is discharged. The initial γ-ray may have a distant astronomical origin, if no local source is present. It has been suggested that a neutron star ceases to be a pulsar when its accelerating electric field drops below the minimum necessary for vacuum breakdown by this process.

A rough dimensional argument makes it possible to estimate the torque on the star. Near the stellar surface the charge density ρ_e (4.4.19), combined with relativistic acceleration, implies a limiting current density $\sim \rho_e c$ which can be accelerated electrostatically. The area of surface through which this current density flows is $\sim R^2\theta_0^2$, implying a total current

$$J \sim B_0 R^2 \Omega \theta_0^2 \\ \sim B_0 R^3 \Omega^2 / c. \tag{4.4.21}$$

This current exerts a force on the star (and *vice versa*) of $\vec{J} \times \vec{B}/c$ per unit length. It extends a length $\sim R\theta_0$ under the polar cap, and is a similar distance from the rotation axis. The magnitude of the resulting torque is

$$N \sim \frac{JB_0}{c}(R\theta_0)^2 \\ \sim \frac{B_0^2 R^6 \Omega^3}{c^3}. \tag{4.4.22}$$

This torque leads to spin-down at the rate

$$\dot{\Omega} \sim -\frac{B_0^2 R^6 \Omega^3}{c^3 I}, \tag{4.4.23}$$

which is comparable to the vacuum radiation torque on an oblique dipole moment (4.4.6). The numerical coefficients in these expressions depend on the detailed electrodynamic structure of the magnetosphere, and in particular on the supply of charged particles, and may change with time. They are usually assumed to be of order unity, but this is not proven, and could be very much less for a pulsar whose magnetosphere is starved of particles. The vacuum torque may be estimated by a related argument, by considering the radiative boundary conditions at the speed of light cylinder.

The power required to accelerate the charged particles is drawn from the neutron star's rotational energy, a configuration known as a unipolar (or homopolar) generator. The circuit potential is the power divided by the current $\Delta\Phi \sim N\Omega/J$, which agrees with (4.4.20). The impedance $\Delta\Phi/J \sim 1/c$, which is characteristic of free space and radiation boundary conditions. If the pulsar is starved of particles J is much less and the impedance is much greater. Most of the potential drop may be across a vacuum gap akin to the brushes of a conventional generator, and which may limit the flow of current.

4.5 Gamma-Ray Bursts

4.5.1 Phenomenology Gamma-ray bursts are rare and unpredictable events. Like pulsars, they were unanticipated, and were found when a new class of instrument, designed to perform another novel task, began to collect data. Gamma-ray bursts were discovered (Klebesadel, *et al.* 1973) by satellite-borne instruments designed to search for a different rare and unpredictable event—clandestine nuclear explosions in space, none of which have ever been observed. Such instruments must be sensitive at all times to radiation arriving from a large fraction of the celestial sphere. Because γ-rays are penetrating it is possible to build such an instrument, which does not focus or collimate the radiation, but only measures the energy deposited in an absorbing medium (a scintillator or electronic detector). Several symposia (Lingenfelter, *et al.* 1982; Burns, *et al.* 1983; Woosley 1984) review observations and theory of γ-ray bursts.

Gamma-ray bursts are a very diverse class of events. They are observed as sources of fairly low energy γ-rays, with most of their energy typically between 100 KeV and 1 MeV, although in many cases photons are detected at energies up to $\sim$ 10 MeV. These photon energies are orders of magnitude lower that those at which some accreting neutron stars (**4.2.4**) and pulsars (**4.4**) are observed. Burst spectra are usually featureless, often roughly (but not closely) approximating power laws. In a few cases spectral features have been reported, either at photon energies of tens of KeV, where they have been suggested to be electron cyclotron resonance lines in strong ($10^{12} - 10^{13}$ gauss) magnetic fields, or near 400 – 500 KeV, where they may result from positron annihilation, possibly redshifted by a neutron star's gravitational field. The interpretation of these reported features is controversial.

A typical γ-ray burst lasts for tens of seconds, but observed durations range from ~ 0.1 second to several minutes. During this period its intensity may fluctuate erratically, with many narrow peaks whose width is apparently unresolved by existing instruments.

The directions from which γ-ray bursts come appear to be isotropically distributed on the sky. Because their detectors have a nearly isotropic response, the direction of the source of radiation can be determined only by comparing the times of arrival of the γ-rays at several detectors distributed throughout the Solar System. The brief duration and rapid fluctuations of γ-ray bursts make this possible, but for only a few bursts is an accurate position known. Studies of the few hundred observed bursts, for most of which positional information is only rough, show that (with a very few exceptions) they are not correlated or associated with any other class of astronomical object. With similarly few exceptions, there is no evidence of individual sources emitting bursts more than once in the nearly twenty years of data now available.

Perhaps ~ 10 γ-ray bursts are observed each year with fluence (the integral of the flux over time) $\sim 10^{-4}$ erg/cm^2. This corresponds to a mean flux passing through the Galactic disc of $\sim 10^{-11}$ erg/cm^{2}sec. If this flux passes through the entire disc the mean Galactic luminosity in γ-ray bursts is $\sim 10^{35}$ erg/sec, a result which is not sensitive to the very uncertain distance and luminosity of individual bursts.

The Galactic γ-ray burst luminosity is remarkably small. The corresponding stellar visible luminosity is $\sim 10^{44}$ erg/sec, that in X-rays is $\sim 10^{40}$ erg/sec, and that in cosmic rays $\gtrsim 10^{41}$ erg/sec. The low mean luminosity of γ-ray bursts may make them harder to understand, because many possible processes will be able to supply the required power, including rare variants of more familiar processes, and because the incidental effects of the release of such little energy are not likely to be conspicuous clues.

These problems are compounded by the scarcity of accurate coordinates on the sky. Not only is it impossible to anticipate a γ-ray burst in order to watch its source in some other wavelength band, but in only a few cases has it been possible even to examine afterwards the direction whence it came.

The distances of γ-ray bursts are unknown. In principle, the distribution of their directions indicates their spatial distribution within the Galaxy, supposing their sources to be Galactic and distributed in a manner similar to that of some more familiar class of objects (if they are extragalactic their luminosities are much higher, and the problems of explaining them more severe). There is not yet evidence for anisotropy in their distribution on the sky, and therefore no evidence for spatial structure resembling the Galactic disc. If it is nonetheless assumed that the sources of γ-ray bursts are distributed like the stars in the Galactic disc, then the failure to observe anisotropy imples that the sources of the observed bursts are closer to us ($\lesssim$ 100 parsecs) than the thickness of the Galactic disc. Such nearby sources would produce an isotropic distribution of arrival directions, even though the more distant sources are anisotropically distributed. The data are equally consistent with any other geometry in which the sources of the observed bursts are distributed isotropically around us, such as a very large halo enveloping the entire Galaxy, or uniform distribution throughout the Universe.

In most populations of astronomical objects the fainter members are much more numerous than the brighter ones, with the number N observed above a threshold S varying as

$$N \propto S^{-d/2} \tag{4.5.1}$$

if the objects are distributed homogeneously throughout a space of d dimensions. For a uniform distribution in space $d = 3$, in a flat sheet $d = 2$, and in a line $d = 1$. This scaling applies to any quantity S which has an inverse square law dependence on distance, and follows directly from the fact that the volume contained within a distance r varies $\propto r^d$. S is typically a flux density or intensity of radiation, but could be the fluence. For γ-ray bursts it is a more complicated function of these quantities, because instrumental thresholds may depend on a complex manner on the variation of the flux with time. The dependence of N on S for the fainter γ-ray bursts implies $d \lesssim 2$, which can be explained only if their sources are distributed in a flat sheet (Galactic disc) or even a line (spiral arm). Such small values of d are obtained only if typical distances exceed the thickness of the Galactic disc (or spiral arm), which would contradict the observation of isotropy. This discrepancy depends only on geometry, and not on any models of the physics of γ-ray bursts.

There are at least two distinct possible resolutions of the discrepancy between the isotropy and the flux distribution of γ-ray bursts. Conceivably the experiments are undercounting weak bursts, or failing to detect an actual anisotropy. It is unfair and probably unjustified for a theoretician (like myself) to suggest that experimenters do not understand their instruments. Alternatively, the sources may be distributed through a volume of finite extent, with most bursts anywhere in this volume luminous enough to be observed. The limit in which all bursts are observed above a threshold S_0 corresponds to $d = 0$ (a point distribution) in (4.5.1). To explain the observed isotropy we must be near the center of this volume (a Ptolemaic γ-ray burst universe; see Cline in Woosley 1984). If this is correct, the sources must either be at cosmological distances, or be distributed in an extended halo about our Galaxy, or be evidence of a previously unsuspected Earth-centered spherical distribution. Of these possibilities, a Galactic halo is the least implausible. The halo's radius must be several times our distance ($\approx$ 10,000 parsecs $\approx 3 \times 10^{22}$ cm) from the Galactic center. It is believed (based on dynamical measurements) that the mass distributions of many galaxies similarly extend far beyond their visible radii. It must then also be that γ-ray burst statistics are not dominated by nearby events of low luminosity, which would tend to produce a $d = 3$ distribution of S.

In a very few cases additional facts are known about individual γ-ray bursts for which accurate coordinates have permitted further study. Schaefer (1981) discovered that in a few cases the source of a γ-ray burst is also the source of a brief flash of visible light, typically with $\sim 10^{-3}$ of the fluence of a γ-ray burst. He made this discovery by examining the extensive archives of photographs obtained with small telescopes for the long-term monitoring of variable stars, which are the classical astronomical data closest (but not close) to a continuous monitoring of the sky. Because these photographs were largely obtained prior to the beginning of γ-ray observations, there was no direct evidence that the γ-ray and visible burst emission were in fact simultaneous. A given γ-ray burst source produces visible flashes at a rate of ~ 1/year (Schaefer and Cline 1985). This is apparently higher than the repetition rate of γ-ray bursts, but each rate is dependent on the detection threshold in its spectral region, and therefore need not be the same if bursts from a given source vary in intensity.

A γ-ray burst observed on March 5, 1979 is famous for several reasons (Cline 1980). Its intensity was very high at its peak, rose extraordinarily rapidly, and later oscillated periodically. Observations by instruments on nine separate spacecraft produced an abundance of data, and provided accurate coordinates on the sky. Its position coincides with that of a young supernova remnant located in the Large Magellanic Cloud (LMC), a small galaxy which is a satellite of our own.

It is not possible to make an unbiased estimate of the statistical significance of this positional coincidence, because the hypothesis to be tested was only defined after the coincidence was noted. Many examples of spuriously significant positional coincidences exist in astronomy (Field, *et al.* 1973). If this bias is ignored the identification of the γ-ray burst with the supernova remnant appears unlikely to be accidental. Inspection of a photograph of the sky with the burst's position lying atop the small image of the supernova remnant has persuaded many, but not all, astrophysicists that they must be associated. If accepted, this association provides the only measurement of the distance to a γ-ray burst, because the distance to the LMC is well determined by classical astronomical methods (studies of pulsating variable stars with calibrated luminosities) to be 55,000 parsecs. From this distance estimate, an assumption of isotropic emission, and the observed flux and fluence, the energy released in the burst is calculated to be $\sim 4 \times 10^{44}$ erg, and the peak luminosity, which lasted ~ 0.1 second, was $\sim 5 \times 10^{44}$ erg/sec (about 10 times the luminosity of the whole Galaxy!).

The γ-ray burst of March 5, 1979 had a number of other interesting properties. Because it was identified with a supernova remnant, it was natural to suggest that it was produced by a compact remnant of stellar collapse, presumably a neutron star. Although its flux had a very brief peak, it was observed for several minutes afterward, and clearly shows a regular variation with a period of 8 seconds. In analogy to pulsars and accreting neutron stars, this is interpreted as the rotation period of a magnetized neutron star, whose intensity of radiation or direction of emission varies with azimuthal angle about its rotation axis. The inference of a neutron star is further supported by the observation of a spectral feature at a photon energy of 420 KeV, which is consistent with the 511 KeV photons produced by the annihilation of positrons, redshifted by the gravitational potential of a neutron star. The radiated intensity rose extremely rapidly, in a time less than the experimental resolution of 2×10^{-4} second, implying (1.6.7) an emission region whose size was $\lesssim 6 \times 10^{6}$ cm, again indicating a neutron star (or black hole, but this cannot explain the other observations).

Is the March 5, 1979 γ-ray burst representative of γ-ray bursts in general? There are obvious differences and unique features in the time dependence of its flux and in its spectrum, but some of these may be attributed to the fact that its great intensity and the large number of functioning detectors made it unusually well studied. Because bursts are such a heterogeneous class of events, its observed properties may not be any more atypical than those of many other bursts. If most other bursts are at distances less than the thickness of the Galactic disc then the March 5, 1979 event was extraordinary, because its luminosity would then be $\sim 10^5 - 10^6$ times greater than that of a typical burst. It is hard to reconcile this interpretation with the absence of bursts similar to that of March 5, 1979 but occurring in our own Galaxy, which contains ~ 10 times the mass of the LMC, and might therefore be expected to have ~ 10 times as many such events. If, instead, bursts in general are distributed throughout a very large halo of our Galaxy, extending to distances comparable to that of the LMC, then the March 5, 1979 burst would be explicable as an unusually (but not extraordinarily) luminous member of a class of events with a broad distribution of luminosities.

4.5.2 Models There is no agreement on even the outline of a model of γ-ray bursts, beyond a consensus that most of the evidence points to magnetic neutron stars. Because their distances are unknown, the luminosity which must be explained is uncertain by several orders of magnitude, ranging from $\sim 10^{37}$ erg/sec for a typical burst within the Galactic disc to $\sim 10^{45}$ erg/sec for the peak luminosity of the burst attributed to the LMC. A distribution throughout a Galactic halo would be consistent with the high space velocities of pulsars (**4.4**) if most γ-ray burst sources were pulsars $\gtrsim 10^8$ years ago (the time required to fill the halo from the birthplace of pulsars in the Galactic disc). Such a halo distribution would imply that most bursts have luminosities and distances approaching those of the burst of March 5, 1979, if its source was in the LMC.

The first task of the modeller is to determine the physical processes which produce the observed radiation. The spectrum of the γ-rays is clearly not that of black body radiation, so it provides evidence of the spectral emissivity of the emission process. Bremsstrahlung (**2.6.1**) and synchrotron radiation (**2.6.2**) by moderately relativistic electrons ($E \sim 1 - 10$ MeV) in neutron star magnetic fields have been suggested, and each is capable of approximating the observed spectrum. In intense magnetic fields and low densities of matter (conditions expected near neutron stars) synchrotron radiation is a much more rapid process, and it is a more popular explanation. Its chief difficulty is that the process which accelerates the radiating electrons (or positrons) must impart to them a component of momentum perpendicular to the magnetic field, which is very hard to do in intense fields. Curvature radiation by very energetic particles moving parallel to the field lines has been considered, as for pulsars, and by producing pairs in the field it may also be a source of moderately relativistic particles with transverse momenta which may radiate synchrotron radiation.

The reported annihilation line radiation poses special problems. The observation of a single gravitational redshift implies that it is produced at a neutron star surface, rather than throughout a magnetosphere. The fact that the line is narrow enough even to be recognized as such implies that it is produced by positrons stopped in relatively cool matter (so that the Doppler broadening is small). It also must not have undergone Compton scattering, which would broaden it, an argument which places an upper bound on the integral of the electron density along paths between the annihilation line source and the observer. Combining this condition with the density required to produce the inferred annihilation rate implies that the line must originate in thin dense sheets or layers of pair plasma, perhaps resembling the auroral sheets produced when fast charged particles enter the Earth's atmosphere from its magnetosphere.

If the distances to γ-ray bursts are more than the thickness of the Galactic disc (as would be true by a large factor for a burst in the LMC), then their luminosities exceed the Eddington limit (1.11.6), and the outward force of radiation pressure on their plasma exceeds the attraction of gravity. If the radiating plasma is composed of electron-positron pairs rather than electrons and ions then the electron scattering opacity (per gram) is greater than that of ordinary matter by a factor $\sim m_p/m_e$, and L_E becomes

$$\begin{aligned} L_0 &= \frac{4\pi c G M m_e}{\sigma_{es}} \\ &= \frac{3GMm_e^3c^5}{2e^4} \\ &= 7 \times 10^{34} \frac{M}{M_\odot} \text{erg/sec}, \end{aligned} \tag{4.5.2}$$

which almost certainly exceeds the luminosity of γ-ray bursts. This limit need not constrain the luminosity of γ-ray bursts because the magnetic field of a neutron star, anchored to currents within its interior, is strong enough to confine the radiating plasma.

The recorded visible outbursts from the sources of γ-ray bursts pose yet another problem. It is natural to ascribe them to radiation by a binary companion star, a familiar phenomenon in high energy astrophysics. This star's surface would be illuminated and heated by γ-rays, and would reradiate its absorbed energy in visible and ultraviolet light, a process which is known from binary X-ray sources. Searches for companion stars when the source is not radiating γ-rays have shown that such stars, if present, must be very faint. If the γ-ray burst sources are close (within the Galactic disc), this establishes that any companion stars must have such low luminosity that they cannot be burning hydrogen. If the bursts come from a Galactic halo a companion star would be much harder to detect, but the inferred visible luminosity during outburst would be very high.

The ratio of the visible luminosity during outburst to that in the dormant state is at least 10^8, which is hard to explain as the consequence of the heating of any black body which has a stellar surface temperature when dormant. In one alternative model the visible radiation is produced nonthermally by energetic particles within the pulsar magnetosphere (synchrotron radiation, curvature radiation, and plasma processes are each possible explanations). There is no direct evidence that the visible and γ-ray bursts are actually simultaneous, and in another alternative model the visible bursts are produced by a binary companion independently of the γ-ray bursts.

The observation of photons with energy $E \gtrsim 1$ MeV is a problem if the burst sources are as distant as the Galactic halo or the LMC. Two photons may produce an electron-positron pair

$$\gamma + \gamma \rightarrow e^+ + e^- \tag{4.5.3}$$

if the photon energies E_1 and E_2 satisfy the kinematic threshold condition

$$\left(\frac{E_1}{m_e c^2}\right)\left(\frac{E_2}{m_e c^2}\right) \geq \frac{2}{1 - \cos\theta}, \tag{4.5.4}$$

where θ is the angle between the γ-ray directions. The cross-section $\sigma_{\gamma\gamma}$ has a peak value of a few tenths of σ_{es}, which is achieved if the threshold condition is exceeded by a factor ~ 2. Near the surface of a body of radius R which radiates isotropically the γ-ray directions will be distributed over 2π steradians. The requirement that γ-rays with $E \gtrsim 1$ MeV escape freely implies that their density n_γ satisfy

$$n_\gamma \sigma_{\gamma\gamma} R \lesssim 1. \tag{4.5.5}$$

This condition may be rewritten in terms of the luminosity in γ-rays of this energy $L_\gamma \sim 4\pi R^2 m_e c^2 n_\gamma c/3$:

$$\begin{aligned} L_\gamma &\lesssim \frac{R m_e^3 c^7}{e^4} \\ &\lesssim \left(\frac{Rc^2}{GM}\right) L_0, \end{aligned} \tag{4.5.6}$$

where L_0 is defined in (4.5.2). The factor in parentheses is about 5 for neutron stars.

This limit on the luminosity of a source of MeV γ-rays is numerically much smaller than L_E, and has been used to argue that the sources of γ-ray bursts must be at relatively small distances (typically $\lesssim 100$ parsecs, but depending on assumptions or measurements of the source geometry, the shape of the spectrum, and the angular distribution of radiation). It cannot be avoided by invoking a confining magnetic field, but can be in other ways. The γ-rays lost by (4.5.3) will be recovered when the pairs annihilate, although repeated cycles of pair production and annihilation will gradually shift the spectrum to lower photon energies because of the influence of processes like $e^+ + e^- \rightarrow \gamma + \gamma + \gamma$ and $\gamma + \gamma \rightarrow e^+ + e^- + \gamma$.

The pair production threshold (4.5.4) is very high for collimated radiation, which may be produced in several ways. Radiation emitted by a spherical surface becomes geometrically collimated as it flows outward. Even a few radii from the source all rays are confined within a narrow bundle, and the threshold (4.5.4) is much higher than at the surface. It is possible that the γ-ray burst source region has a complex structure, with the lower energy γ-rays produced near the neutron star, and the more energetic γ-rays escaping only if produced at higher altitudes, and directed outwards. The radiation of collimated relativistic particles, such as the curvature radiation of particles moving outward on magnetic field lines, is intrinsically collimated. The bound (4.5.6) should therefore not be uncritically assumed to constrain γ-ray burst luminosities and distances, although the construction of models which exceed it by large factors would require a detailed knowledge of the particle acceleration and radiation processes throughout the magnetosphere.

The source of energy for γ-ray bursts is also unknown. Three models have been investigated. In the first model the energy is accretional, and its sudden release is attributed to the infall of a solid object (comet or asteroid) onto a neutron star, or to an instability in the accretion of plasma. The strength of this model is that accretion certainly takes place, although the rate of accretion of solid bodies is uncertain and probably low. Its weakness is that we observe accretion onto neutron stars in the luminous X-ray sources (**4.2.1**), where it produces X-rays of $\sim 1-30$ KeV energy, quite unlike the spectrum observed for γ-ray bursts. Matter suddenly accreted onto a neutron star is unlikely to be confined by a magnetic field, having just fallen from outside, so that the emergent luminosity should satisfy $L < L_E$. This is consistent with the fluxes observed from γ-ray bursts if their sources are within the Galactic disc, but not if they are distributed throughout a Galactic halo. The rapid rise of the flux of the March 5, 1979 burst leads to a similar conclusion, even if no assumption is made about its mass. If its energy was accretional (4.7.1) implies that $L \lesssim 10^{39}$ erg/sec, requiring a distance $\lesssim$ 100 parsecs, well within the Galactic disc rather than in the LMC.

A second model uses the thermonuclear energy contained in matter accreted onto neutron stars. This energy is available, because matter is known to accrete onto neutron stars and calculations indicate that nuclear burning will occur and will produce thermonuclear runaways. The difficulty with these models is that thermonuclear energy release on neutron stars occurs at high optical depths, so that the power will be radiated from the surface with roughly a black body spectrum. This is a satisfactory model of X-ray bursts (**4.2.3**), but therefore has difficulty accounting for the very different spectrum observed for γ-ray bursts. It would predict a smooth variation of luminosity with time in a burst, with only one or two maxima, rather than the very irregular behavior observed. It has difficulty explaining the apparent frequency (Schaefer and Cline 1985) of visible bursts from γ-ray burst sources without evidence of a binary companion to supply a high rate of accretion. It predicts a mean accretional X-ray luminosity $\sim$ 100 times larger than the mean γ-ray burst luminosity, which has not been observed.

A third model describes γ-ray bursts as neutron star analogues of Solar flares, releasing magnetostatic energy in optically thin regions above the surface and accelerating relativistic particles there. This model is suggested by the observed irregular variation of luminosity with time within a burst, and the extremely rapid rise sometimes observed. It is consistent with the observed spectrum, and may provide a nonthermal explanation of the optical bursts, although the model is not well enough understood to have predicted either. Its chief difficulty is that it is not possible to say why, when, or where such flares should take place. It is also difficult to store enough energy in a neutron star's magnetosphere, and to convert it suddenly and efficiently to the acceleration of particles of the required energy. It may be necessary to regenerate the magnetospheric energy, perhaps from the differential rotation of a neutron star's superfluid interior, but there is little evidence for the existence of a significant reservoir of energy. There is also no observed correlation between γ-ray bursts and pulsar glitches, which in this model might be related events.

The observations which offer the most promise of clarifying this muddy picture are the monitoring of the sky for fast visible transients (see articles by Ricker, *et al.* and Teegarden, *et al.* in Woosley 1984). These observations may also aid the understanding of novae and supernovae, as well as other transient astronomical events. The simultaneous observation of a visible and a γ-ray burst will immediately decide whether the visible radiation is γ-ray energy absorbed and reradiated by a binary companion star, and will thus demonstrate or exclude the presence of such a star as a source of matter for accretion. Nonthermal visible radiation would have a time dependence very different from that of reradiation (which would be the convolution of the γ-ray power with a geometrical time delay and the thermal response of the heated stellar atmosphere), and might also be strongly linearly polarized. If the visible burst is reradiation, comparison of the visible and γ-ray bursts will provide detailed geometrical information about the binary system.

4.6 Accreting Black Holes

Black holes have been the subject of extensive research since the early days of general relativity, and their properties have been predicted in great detail. With no nearby black hole available for experimentation it has not been possible to test these predictions, so relativists have watched with great interest the astronomical search for black holes. Isolated black holes are expected to be nearly undetectable, and most interest has concentrated on accreting black holes. These should be ideal examples of accretion flows (**3.6, 3.7**), because there is no stellar surface to add its own contribution to the radiation, or stellar magnetic field to interfere with the flow. Unfortunately, this means that an accreting black hole has no predicted characteristic signature (unlike the rotation period of a magnetic neutron star, or the X-ray bursts emitted by thermonuclear energy release on other neutron stars). The properties of its radiation are largely determined by the hydrodynamics of the accretion flow rather than any exotic relativistic properties of the black hole. This has disappointed relativists in two ways: it has been difficult even to identify black holes, and once identified observations have provided information about accretion hydrodynamics rather than about relativity.

Because they have no characteristic signature, all accepted inferences of black holes have been based on a measurement of the mass of a candidate object in a binary stellar system. These candidates are usually luminous sources of X-rays, which are attributed to an accretion flow into a deep gravitational potential well. In order to rule out a neutron star or a degenerate dwarf the mass, determined from Newton's laws applied to the orbital motion of the companion star, must exceed M_{Ch} and M_{Ch}^{ns} (**1.12**). Because $M_{Ch} = 1.40 M_\odot$ and M_{Ch}^{ns} is believed with some confidence to be $< 3M_\odot$, the generally accepted criterion for the identification of a black hole is the observation of a compact object whose mass safely exceeds $3M_\odot$.

A black hole which is not accreting will not attract attention to itself as an X-ray source. The existence of such a black hole could be inferred from the orbital motion of a binary companion star, but without X-ray emission it also would be necessary to rule out the possibility of its being an ordinary massive star. Such stars are generally very luminous (**1.11**), and thus easy to exclude if their expected luminosity is not observed, but a skeptic might suggest that the star is present but subluminous because of peculiarities in its evolution (not implausible in a close binary system), or because it is cloaked by dust. At present, no such identifications of non-accreting black holes are generally accepted.

The first, and for many years the only, accepted black hole identification was the X-ray source Cygnus X-1. Its X-ray properties are very complex. Two distinct states are observed, with different spectra and luminosities. Simple theories of disc accretion (**3.6**) predict that the characteristic temperature (3.6.12) of the inner parts of a disc radiating as a black body should vary $\propto \dot{M}^{1/4} M^{-1/2}$, so that an accreting black hole would be predicted to have a somewhat softer (lower photon energy) spectrum than that of an accreting neutron star of comparable accretion rate and luminosity. The X-ray spectrum of Cygnus X-1 is instead observed to be much harder, with abundant radiation at photon energies ~ 100 KeV. The simple theory describes a steady disc, but the flux from Cygnus X-1 is observed to vary substantially in much less than a second. Accretion onto Cygnus X-1 probably takes place from a wind (produced by its luminous binary companion star) rather than by Roche lobe overflow, and it is possible that there is no disc. Alternatively, the direction of disc angular momentum may vary, which will affect its properties if the black hole also has angular momentum.

Shapiro, *et al.* (1976) have argued that the observations of Cygnus X-1 are consistent with more complex models of accretion discs, in which the emitting matter is very hot and optically thin. These models can account for many of its observed properties, but it is not clear why some accretion discs should have this structure while others do not.

It has been suggested that very massive black holes might be found at the centers of globular clusters, galaxies, and quasars. The masses suggested have ranged from $\sim 10^3 M_\odot$ to $\sim 10^{10} M_\odot$, very roughly in proportion to the mass of the entire system (in order to avoid large disturbances to the stellar orbits in the cluster or galaxy, the black hole can have only a small fraction of its total mass). These black holes might be associated with the formation of these systems, which is not understood, or produced in their later evolution (**2.4**).

It is now known that the source of the X-ray emission observed from some globular clusters is not a central black hole because the X-rays do not come from the cluster centers (Grindlay, *et al.* 1984). Accretion onto very massive black holes would be expected to produce visible or ultraviolet radiation, rather than X-rays. In a popular model quasars (**4.7**) are the consequence of accretion onto a black hole at the center of a galaxy. The thermal radiation of an accretion disc may contribute to the blue and ultraviolet emission from some quasars, but it is not the principal source of their radiation.

It may be possible to detect massive central black holes by their influence on the distribution and motions of stars and gas around them, but claims to infer the presence of black holes on this basis remain controversial. It is difficult in practice to identify the effects of a black hole on a distribution of stars if the black hole contains only a small fraction of the total mass. Such a black hole dominates the gravitational potential in only a very small fraction of the total volume, and a dense (but far from relativistic) inner nucleus of stars may resemble the central concentration produced by a black hole. Observations with much improved spatial resolution are required to solve this problem.

Black holes are perhaps the most exotic and intriguing objects in astrophysics. Disappointingly, the study of their accretion has contributed so little to their understanding or to that of accretion flows that it is not possible even to distinguish them from neutron stars except by measuring their masses.

4.7 Quasars

The study of quasars involves many different problems and phenomena, ranging from the source of their power to the properties of the gas which produces their observed atomic line radiation. The field is so broad that there is no recent comprehensive review, but a symposium (Ulfbeck 1978) provides a broad survey.

The definition of a *quasar* (a name derived from the more cumbersome expressions quasi-stellar object or quasi-stellar radio source) depends on the astronomer, but usually includes the condition that the image on a particular series of photographs (the Palomar Sky Survey, taken with a 48 inch Schmidt telescope) resemble that of a star. This definition is clearly dependent on the instrument and observing conditions. Better photographs often show distinctly extended images, and the Space Telescope may show a great deal of structure. It is unfortunate that the balloon-borne Stratoscope telescope could not obtain images of quasars, although it did obtain an image of a related active galactic nucleus (Schwarzschild 1973; Light, *et al.* 1974). This telescope preceded space telescopes by more than 15 years. Larger balloon-borne telescopes, equipped with modern detectors, may be capable of angular resolutions equal to those of space telescopes, with greater flexibility and at lesser cost.

Quasar spectra contain strong emission lines (though there exist BL Lacertae objects which resemble quasars except that they show no spectral lines at all). This line radiation (**1.14**) is probably produced by ions ionized or excited by the quasar's ultraviolet continuum radiation. The emission lines are observed to have large redshifts, which are attributed to the expansion of the universe and which imply that quasars are very distant and very luminous.

The characteristic observational properties of quasars have been known since their recognition in the mid-1960's: luminosity, variability, and spectrum. The visible luminosity of a quasar may be $\sim 10^{46} - 10^{47}$ erg/sec, more than 100 times the luminosity of a bright galaxy. In some cases it varies on a time scale of days. This variability, a power-law spectrum of the continuum radiation, and linear polarization of several percent in some quasars are evidence that much of their radiation has a nonthermal origin. Many quasars are also observed to be radio and X-ray sources. Their radio emission is certainly synchrotron radiation by relativistic electrons, and their X-ray emission is probably also nonthermal.

Quasars are observed at higher redshifts than any other discrete astronomical objects, with values of redshift $z \gtrsim 2$ very common ($z \equiv \Delta\lambda/\lambda_0$, where λ_0 is the wavelength at rest). Ultraviolet lines like Lyman α (1216 Å) are shifted into the visible spectrum. At this distance an ordinary galaxy would be nearly undetectable, and if detected would have an image barely larger than that of a star seen through the Earth's atmosphere. The slightly extended images observed for many quasars are widely believed to be luminous galaxies at whose centers the quasars are found. When images and spectra of sufficient spatial resolution to separate cleanly the extended image from the central quasar are obtained, this hypothesis will be tested. It could instead be that these extended sources of radiation are largely gas excited by the quasar's energy, thus resembling a giant H II region (a gas cloud ionized by ultraviolet radiation from a central source) or a supernova remnant more than an ordinary galaxy.

The most natural and familiar interpretation of quasar continuum radiation is that it is electron synchrotron radiation. Unfortunately, simple synchrotron models are inconsistent with the small sizes inferred (1.6.7) from the rapid variability of some quasars and related objects. The synchrotron radiation energy density is so high that the relativistic electrons lose more energy to Compton scattering than to synchrotron radiation itself. The Compton scattered radiation therefore has a much higher energy density than the original synchrotron radiation field. The power an electron loses to Compton scattering (2.6.36) is proportional to the energy density of the scattering radiation field. The power lost by scattering some photons a second time then far exceeds that lost in scattering the original synchrotron radiation, and the power lost by scattering a few photons a third time exceeds that lost in the second scatterings, ... (Hoyle, *et al.* 1966). The result would be an enormous (though not infinite) energy loss by Compton scattering which is inconsistent with the model of a synchrotron radiation source, and which would imply that most of the quasar's luminosity is emitted as X-rays and γ-rays, in conflict with observation.

This "Compton catastrophe" led some astronomers to question the luminosity estimates, which are based on the interpretation of the redshifts as those of cosmological expansion. Unfortunately, no satisfactory alternative explanation of the redshift has ever been suggested, and strong circumstantial evidence supports the cosmological interpretation. A long and fruitless controversy has raged concerning the estimates of quasar luminosity and the interpretation of their redshifts (Field, *et al.* 1973). Nearly all astronomers accept the redshifts as cosmological, but a few skeptics remain.

Ingenious theorists have suggested ways to avoid the "Compton catastrophe." The most plausible involves an anisotropic distribution of relativistic electrons. If they are directed nearly radially outward, as would be expected if they are produced in and stream away from a central source, then their synchrotron radiation would be similarly directed outward (by the kinematics of relativistic particle radiation). The radiation energy density in the electrons' frame is then much less than that in the laboratory frame, reducing the rate of Compton scattering (Woltjer 1966). This natural and attractive mechanism can produce an arbitrarily large reduction in the rate of Compton scattering, and thus can eliminate the "Compton catastrophe," if the electron collimation is good. Observations of superluminal expansion (a size which appears to grow at a speed faster than c) in many radio sources similarly imply the directed outflow of relativistic electrons (**1.6**).

In an historical review Weedman (1976) argued that in slightly different circumstances quasars would never have been classified as a distinct class of objects, but would have been considered extreme examples of previously recognized classes of galaxies (Seyfert and N-galaxies) with peculiar active nuclei. Many of these active galactic nuclei produce variable, polarized, visible emission with power law spectra, and appear to be lower luminosity examples of the same phenomenon observed in quasars. These galaxies are close enough to us that their images look like those of galaxies, and values of their redshifts may be measured from the galactic images (rather than from the active nuclei). The interpretation of these redshifts as those of cosmological expansion has been uncontroversial. Some of these objects are sufficiently luminous and variable (on time scales shorter than those of quasars) that simple synchrotron models lead to the "Compton catastrophe." Had this problem first been discovered for Seyfert galaxies, the redshift controversy might never have arisen. It is likely that quasars are distinguished from their less controversial siblings only by greater distance and more luminous nuclear activity, which obscure their galactic nature and give them a point-like (quasi-stellar) appearance.

It is possible to predict a limit to the rapidity with which a luminous object, producing energy by accretion, can vary. Requiring $L < L_E$ (1.11.6), $t_{var} \gtrsim R/c$ (1.6.7), and writing κ_{es} in terms of physical constants (2.6.30,32), implies

$$\begin{aligned} t_{var} &\gtrsim \left(\frac{Rc^2}{GM}\right) \frac{e^4 L}{m_e^2 m_p c^8} \\ &\gtrsim 6 \times 10^{-44} \left(\frac{Rc^2}{GM}\right) \frac{\text{sec}^2}{\text{erg}} L. \end{aligned} \tag{4.7.1}$$

If the object is a source of γ-rays with energy $E \gtrsim 1$ MeV and luminosity L_γ then (4.5.6) implies

$$\begin{aligned} t_{var\gamma} &\gtrsim \frac{e^4 L_\gamma}{m_e^3 c^8} \\ &\gtrsim 1 \times 10^{-40} \frac{\text{sec}^2}{\text{erg}} L_\gamma. \end{aligned} \tag{4.7.2}$$

The observed variations of quasars and active galactic nuclei satisfy these constraints. Their violation would not contradict any fundamental law of physics, but would require either some other immediate energy source (for example, the release of magnetostatic energy, although the ultimate source could still be gravitational), directed relativistic motion, anisotropic emission, or (4.7.1 only) magnetic confinement.

In the early days of quasar studies it was thought that their high power and compact size (implied by their variability) were inexplicable. This was probably the reaction of astrophysicists startled by the apparent novelty and extraordinary energy scale of the phenomenon. It is now widely accepted that the observed properties of quasars do not violate any physical laws or even other accepted astrophysical ideas, but we are hardly closer to understanding how they actually work.

Quasars' large power and small size point to the release of gravitational energy by accretion into a deep potential well. Many detailed models have been proposed, involving a single very massive black hole, clusters of black holes of stellar mass, of neutron stars, or of pulsars, colliding stars in dense clusters, supermassive stars, supernovae, accretion discs (under a variety of names), and probably every other combination of object (or fantasy) known to the theoretical astrophysicist. Unfortunately, none of these models really explains the required acceleration of energetic particles, or makes other unambiguous predictions. The observed star-like (pseudo-stellar) nuclei of some galaxies radiate ordinary starlight, rather than synchrotron radiation, but may be the ancestors of quasars. They are hints that a galactic nucleus may contain a star cluster dense and massive enough to power a quasar in its death-throes.

Perhaps the most popular model (Rees 1984) is rapid accretion onto a massive black hole. This model is attractively simple, and a massive black hole is a plausible result of the evolution of a dense cluster of stars at the center of a galaxy. Most processes (**2.4**) increase the gravitational binding of such a cluster, and the formation of a black hole may be inevitable. Once formed, it will grow by accretion of gas or even entire stars, at a rate limited only by the supply of mass. The greater uncertainty may be whether massive black holes have formed by the cosmological epoch at which the observed quasars were radiating, rather than whether they form at all.

Black hole accretion is believed capable (as well as these processes are understood) of supplying the required high power, and is consistent with the small size implied by the observed variability. For example, if $M = 10^8 M_\odot$ then $L_E \approx 10^{46}$ erg/sec (1.11.6; see also **3.6, 3.7**), and the characteristic minimum time scale of variability of the inner disc is t_{lt} (1.6.6) or t_{var} (4.7.1), and is $\sim$ 1 hour, consistent with observation.

More difficult is the problem of explaining the acceleration of an enormous flux of relativistic electrons, required to produce the observed radiation by the synchrotron process. Blandford (1976) and Lovelace (1976) have suggested that a magnetized accretion disc may act as a homopolar generator, in analogy to pulsars (**4.4.2**), and have suggested that it may directly convert much of its luminosity to the acceleration of relativistic particles rather than to thermal radiation. These models even provide a natural explanation of the geometry of the jets and double radio sources associated with some quasars and active galactic nuclei. Their difficulty is that in order to obtain particle acceleration the plasma density must not exceed a characteristic value (4.4.19) which is extremely low ($\sim 10^{-10}$ cm^{-3}) for the parameters appropriate to these discs. It is hard to imagine a region of interstellar space, particularly one in which there is a supply of gas to feed an accretion disc, evacuated to this density, which is $\sim 10^{-10}$ of the mean interstellar density.

There are models in which particle acceleration is the consequence of hydrodynamic, magnetohydrodynamic, or plasma turbulence, much as is believed to occur in supernova remnants and the interstellar medium. This turbulence might be excited by energy extracted from an accretion disc electrodynamically, or from other gas motions, such as the expansion of supernova shells. Even if the source of energy is an accretion disc, particles may be accelerated by turbulent processes, rather than by the pulsar process.

In yet another class of models many individual objects of stellar mass supply a quasar's luminosity. Because of the need to accelerate energetic electrons it is natural to assume that these are young rapidly spinning pulsars. Supernovae or stellar collisions might explain the emitted power, but not the efficient acceleration of particles. Power is drawn from the pulsars' rotational kinetic energy, but its ultimate origin is the gravitational binding energy of the neutron stars, whose rotational velocities are multiplied in a collapse which conserves angular momentum. The advantage of this model is that it naturally explains the efficient acceleration of energetic electrons. The bound (4.7.1) also does not apply because the immediate source of energy is not accretional, and $L > L_E$ is possible in a neutron star magnetosphere (L_E must be exceeded by several orders of magnitude).

The multiple pulsar model of quasars would require that rapidly spinning pulsars (spin periods ~ 0.001 second, $B_0 \sim 10^{13}$ gauss) be born several times a year in a dense cluster of rapidly evolving stars. There is no reason, other than the desire to explain quasars, to believe that the evolution of a galactic nucleus will lead to such a cluster.

The observed brightness histories of quasars (Angione and Smith 1985) contain slow trends as well as rapid variations of intensity. Models involving many independent events, such as the birth of pulsars, might be expected not to show slow trends in their total luminosity, but rather to resemble Poisson noise processes. Further, the observed rapid variations in intensity may be hard to explain in a model in which pulsars are born suddenly, but then slow gradually. Ingenious theorists can answer these objections (for example, modulating the emission by immersing the sources in a single gas cloud whose properties vary on both longer and shorter time scales).

None of these models, nor any others, are understood quantitatively, and most models can be bent to fit almost any data. Preferences among them are largely a matter of opinion as to their relative plausibility (or implausibility). It will be very hard to show which model is correct.

4.8 References

Alme, M. L., and Wilson, J. R. 1973, *Ap. J.* **186**, 1015.

Angel, J. R. P. 1978, *Ann. Rev. Astron. Ap.* **16**, 487.

Angione, R. J., and Smith, H. J. 1985, *Astron. J.* **90**, 2474.

Baltrusaitis, R. M., Cassiday, G. L., Cooper, R., Elbert, J. W., Gerhardy, P. R., Loh, E. C., Mizumoto, Y., Sokolsky, P., Sommers, P., and Steck, D. 1985, *Ap. J. (Lett.)* **293**, L69.

Blandford, R. D. 1976, *Mon. Not. Roy. Ast. Soc.* **176**, 465.

Burns, M. L., Harding, A. K., and Ramaty, R., eds. 1983, *Positron-Electron Pairs in Astrophysics* (New York: American Institute of Physics).

Chadwick, P. M., Dowthwaite, J. C., Harrison, A. B., Kirkman, I. W., McComb, T. J. L., Orford, K. J., and Turver, K. E. 1985, *Astron. Ap.* **151**, L1.

Chanmugam, G., and Brecher, K. 1985, *Nature* **313**, 767.

Cline, T. L. 1980, *Comments Ap.* **9**, 13.

Córdova, F. A., and Mason, K. O. 1983, in *Accretion Driven Stellar X-Ray Sources* eds. W. H. G. Lewin and E. P. J. van den Heuvel (Cambridge: Cambridge University Press), p. 147.

Dowthwaite, J. C., Harrison A. B., Kirkman, I. W., Macrae, H. J., Orford, K. J., Turver, K. E., and Walmsley, M. 1984, *Nature* **309**, 691.

Eichler, D., and Vestrand, W. T. 1984, *Nature* **307**, 613.

Field, G. B., Arp, H. C., and Bahcall, J. N. 1973, *The Redshift Controversy* (Menlo Park: Benjamin/Cummings).

Gallagher, J. S., and Starrfield, S. 1978, *Ann. Rev. Astron. Ap.* **16**, 171.

Grindlay, J. E., Hertz, P., Steiner, J. E., Murray, S. S., and Lightman, A. P. 1984, *Ap. J. (Lett.)* **282**, L13.

Gold, T. 1968, *Nature* **218**, 731.

Goldreich, P., and Julian, W. H. 1969, *Ap. J.* **157**, 869.

Hewish, A., Bell, S. J., Pilkington, J. D. H., Scott, P. F., and Collins, R. A. 1968, *Nature* **217**, 709.

Hoyle, F., Burbidge, G. R., and Sargent, W. L. W. 1966, *Nature* **209**, 751.

Joss, P. C., and Rappaport, S. A. 1984, *Ann. Rev. Astron. Ap.* **22**, 537.

Klebesadel, R. W., Strong, I. B., and Olson, R. A. 1973, *Ap. J. (Lett.)* **182**, L85.

Light, E. S., Danielson, R. E., and Schwarzschild, M. 1974, *Ap. J.* **194**, 257.

Lingenfelter, R. E., Hudson, H. S., and Worrall, D. M., eds. 1982, *Gamma Ray Transients and Related Astrophysical Phenomena* (New York: American Institute of Physics).

Lovelace, R. V. E. 1976, *Nature* **262**, 649.

Lyne, A. G., Anderson, B., and Salter, M. J. 1982, *Mon. Not. Roy. Ast. Soc.* **201**, 503.

Manchester, R. M., Durdin, J. M., and Newton, L. M. 1985, *Nature* **313**, 374.

Michel, F. C. 1982, *Rev. Mod. Phys.* **54**, 1.

Protheroe, R. J., Clay, R. W., and Gerhardy, P. R. 1984, *Ap. J. (Lett.)* **280**, L47.

Protheroe, R. J., and Clay, R. W. 1985, *Nature* **315**, 205.

Rees, M. J. 1984, *Ann. Rev. Astron. Ap.* **22**, 471.

Ruderman, M. A., and Sutherland, P. G. 1975, *Ap. J.* **196**, 51.

Samorski, M., and Stamm, W. 1983, *Ap. J. (Lett.)* **268**, L17.

Schaefer, B. E., 1981, *Nature* **294**, 722.

Schaefer, B. E., and Cline, T. L. 1985, *Ap. J.* **289**, 490.

Schwarzschild, M. 1973, *Ap. J.* **182**, 357.

Shapiro, S. L., Lightman, A. P., and Eardley, D. M. 1976, *Ap. J.* **204**, 187.

Shapiro, S. L., and Teukolsky, S. A. 1983, *Black Holes, White Dwarfs, and Neutron Stars* (New York: Wiley).

Sturrock, P. A. 1971, *Ap. J.* **164**, 529.

Trimble, V. 1982, *Rev. Mod. Phys.* **54**, 1183.

Trimble, V. 1983, *Rev. Mod. Phys.* **55**, 511.

Ulfbeck, O., ed. 1978, *Quasars and Active Nuclei of Galaxies* in *Physica Scripta* **17**, 133–385.

Weedman, D. W. 1976, *Quart. J. Roy. Ast. Soc.* **17**, 227.

Woltjer, L. 1966, *Ap. J.* **146**, 597.

Woosley, S. E., ed. 1984, *High Energy Transients in Astrophysics* (New York: American Institute of Physics).

Appendices

A.1 Information in Astronomy

In this appendix I give a brief and rough discussion of the information content, and useful information content, of some astronomical data. The purpose is only to explain why certain kinds of astronomical data have been and are likely to be useful, and others less so; a whiff of information theory is mixed with astronomical experience and cynicism.

If we perform a measurement of a quantity with a signal to noise ratio S/N (or, equivalently, that ratio of measured quantity to its uncertainty) we may express the result as a number in binary notation with $n = \log_2(S/N)$ significant digits. In other language, the result contains n bits of information. n cannot be very large (Fermi is reputed to have said that all logarithms are equal to 10).

If the measured value is zero (or numerically small), as is frequently the case, this definition of n underestimates the significance of the result, and S should be some estimate of the characteristic magnitude the result of our null experiment could have had. For example, if we are measuring the charge of the neutron S might be taken to be the electron charge, even though the experiment has always produced a null result. Clearly we have introduced a subjective element into n. To make it objective and more quantitative would require a careful consideration of the various hypotheses we might consider (and their *a priori* likelihoods). Such exercises in probability theory are rarely justified in experimental science.

We frequently have a series of N similar measurements, representing a time series, its Fourier transform, a spectrogram, an image, or some other array. This array then contains Nn bits of information. Note that a minimum of $2^{2n}N$ photons must have been detected. Values of N are found over an enormous range, from one or a few (classical astronomical photometry) to $10^3 - 10^5$ (optical spectrometry, typical astronomical time series), to 10^9 (good photographic images).

These estimates of information content may offer valuable hints to the opportunities for observing interesting phenomena. For example, most of the interesting (in my narrow opinion!) phenomena in X-ray astronomy have been found in studies of time series and in the spectroscopy of the optical counterparts of X-ray sources; both these kinds of study produce data with fairly large values of N. Astronomical X-ray spectroscopy, for which usually $N \lesssim 100$, has been much less fruitful.

It is necessary, however, to distinguish between information in the sense of the communications engineer, which was the basis of these estimates, and scientifically useful information. If we measure a physical constant N times we should obtain the same value each time, and clearly do not have Nn useful bits of information. Rather, we have n useful bits if the errors are systematic; if they are random and well-behaved statistically we can average our measurements to produce a mean with greater accuracy (up to $n + \frac{1}{2}\log_2 N$ bits) than that of an individual measurement. The extra bits are not lost; rather, we already assumed we knew them when we assumed that we were measuring the same physical constant N times, so that finding them again in the constancy of the data does not tell us anything new.

Images have very large values of N, but have been particularly disappointing in astronomy. One reason is that many astronomical objects (most familiarly, stars) are unresolved in images. Only one independent picture element of the image collects data from the object of interest, and the rest are irrelevant. The second reason is less trivial. It is very hard to formulate quantitative hypotheses to describe usefully the morphology of extended objects, and hence it is hard to utilize the high information content of their images.

Spectrograms have related problems. Wavelengths at which no detectable spectral features exist produce no useful information. Most of the spectral lines observed come from a relatively few species and ionization states, and their intensities are related by inflexible rules of atomic physics. Once we assume the laws of atomic physics to hold everywhere, the number of independent parameters which can be measured is much smaller than N. In X-ray astronomy most spectrograms determine at best a density, a temperature, and a few abundances (in principle the spatial distribution of these quantities is described by an infinity of parameters, but usually the data are sufficient to determine only a very few). The greater spectral resolution and information content of visible spectra permits the resolution of line profiles, and hence the study of additional physics: velocity fields and line-broadening mechanisms.

Time series in astronomy have been very useful when they have led to the discovery of regular periodicity. Time series without such periodicities have been nearly useless, even though in principle they contain a great deal of statistical information, because it has not been possible to formulate scientifically interesting and sensible hypotheses to predict their statistics. For example, a time series of N elements which is observed to be constant (within the statistical errors) contains much fewer than Nn bits of useful information. Instead, it contains no more than $\log_2 M$ bits, where M is the number of scientifically sensible (not *a priori* rejected) hypotheses. Since these hypotheses will generally be expressible as specific functional forms for the signal, with a small number of free parameters, M will be approximately the number of distinct functional forms of hypotheses times 2^{pm} where p is the number of free parameters in each and $m \lesssim n + \frac{1}{2}\log_2 N$ is the number of significant binary bits to which each may be determined. In realistic scientific theories p is a small number and the final result is usually ~ 100 bits, and often less. For a physical constant there is one hypothesis and $p = 1$.

The most familiar and important use of time series is the study of regular periodicities, which are predicted by many theories (as a consequence of orbital, rotational, or vibrational motion), and which are widely observed. A time series of N evenly spaced elements permits a frequency resolving power of $N/2$. It is therefore possible to measure directly a period or frequency stability (quality factor) $Q \equiv 1/|\dot{P}|$ (where P is the period) of approximately $N/2$. This Q may be defined very generally as the width of a Fourier transform, and does not require an assumption of sinusoidal variation. Such a measurement of P or Q contains $\log_2 N$ bits; this is our previous expression with one hypothesis, $p = 1$, and $m = \log_2 N$ (or $M = N$). It has been tacitly assumed that $n \gtrsim \frac{1}{2}\log_2 N$; complications ensue if this is not the case. If exact sinusoidal behavior is assumed the accuracy to which its period may be determined increases to 1 part in $2^n N^{1/2}$, corresponding to $\log_2 M = n + \frac{1}{2}\log_2 N$ bits (note that this accurate determination of a sinusoidal period does not imply that the actual Q of the oscillator is as large as $2^n N^{1/2}$; a much longer time series would be required to establish this). In practice, systematic errors rarely permit this full extra accuracy.

The use of unevenly spaced time series (the only ones available over extended periods when observations are interrupted by daylight, seasonal effects, and other practical problems) permits very much higher accuracy in period determination for a given number of observations than is possible with evenly spaced time series. This accuracy is obtained at the price of failing to distinguish strictly periodic behavior from a large class of aperiodic behavior (periods modulated in regular but complex ways). These aperiodic hypotheses are rejected on sound scientific grounds, but they are fully consistent with the data; these independent scientific arguments add a significant number of essential bits of information.

It is possible to determine a frequency stability Q and to measure the period of oscillation (assumed sinusoidal) to an accuracy of 1 part in Q with $\sim \ln Q$ observations of moderate quality ($n \gtrsim 5$). This method is widely used to measure values of Q as high as 10^{15} in pulsars and other very stable oscillators. A small number j of observations over a period T determine the frequency to an accuracy of about $\pm 1/T$. An additional j observations at intervals of order $T/2$ later extend the time baseline to of order $jT/2$, and the frequency accuracy to about $\pm 2/(jT)$. After k repetitions of this process the time baseline is of order $(j/2)^k T$, the frequency accuracy is of order $2^k/(j^k T)$, and the required number of observations $j(k+1) \approx j \ln Q / \ln(j/2) \gtrsim 2e \ln Q$ varies only logarithmically with Q. Accurate data permit better period determinations in each iteration and longer gaps between subsequent observations; this is required in practice by the timing of opportunities for observation, which are usually brief and widely separated. This reduces the required number of observations by a modest factor.

Another problem in which a qualitative application of these ideas is illuminating is that of deconvolution. Frequently we have data (in the form of a series or array of numbers) which have been smoothed by the limited resolution of an instrument or some other process. Familiar examples are visible images blurred by atmospheric "seeing" and spectrograms smoothed by the finite spectral resolution of the instrument. If the properties of the smoothing function (the atmospheric or instrumental response to a point-like signal) are known then one might hope to remove its influence and to recover the original signals. (Usually the smoothing function is only approximately known, but this is generally not critical.)

A variety of linear and nonlinear algorithms exist to accomplish this process of "deconvolution," but their results are frequently disappointing. The reason is easy to see. Suppose the original observed data consist of U numbers, each known with n significant binary bits, and it is desired to increase the resolution (in one dimension) by a factor q in order to obtain a deconvolved series of $N = qU$ independent numbers. Only nU bits of information are available, so each of the new numbers can only have (on average) n/q significant bits. Because n is rarely more than 10, it is clear that attempts to significantly increase the resolution rapidly destroy the ratio of signal to noise! The most successful algorithms are those which best use the available information to answer the scientifically interesting questions by fitting to an intelligently chosen model, rather than blindly attempting to increase the resolution; those known as "maximum entropy methods" provide a systematic (and in some sense optimal) way of imposing constraints.

It is my opinion that formal statistical methods and tests are frequently misleading and of little real use in experimental science, and that the best way to extract useful scientific information from imperfect data is to compare them to the predictions of sensibly chosen models, using the human eye to assess the results.

A.2 Fermi at Alamogordo

The story is told (I do not vouch for its historical accuracy) that when Fermi witnessed the first nuclear explosion he wanted a quick estimate of the energy Y it released. His observing point was far enough from the explosion (at a distance $R \gg (Y/P_0)^{1/3}$) that its shock wave had become very small in amplitude, with a pressure jump $\Delta P \ll P_0$, where P_0 is the pressure of the ambient air. This is called regime III in **3.4**. When he saw the explosion's flash of light he is said to have dropped a few small scraps of paper into the air. A number of seconds later the shock passed by. Because the scraps of paper were very light they moved with the air, while more massive objects remained fixed. By measuring the displacement ΔR of the scraps of paper he was able to estimate the energy of the hydrodynamic motion produced by the explosion.

If an explosion is close to the surface of the ground (in comparison to the observer's distance from it) its shock will be nearly hemispherical. The mechanical work done by the shock against the pressure of the atmosphere is

$$W = 2\pi R^2 \Delta R P_0. \qquad (A.2.1)$$

Air is well described (under ordinary conditions) by a perfect gas equation of state with a ratio of specific heats $\gamma = c_P/c_V = 1.40$ (**1.9.1**), and $P_0 = 1.0 \times 10^6$ dyne/cm^2. A fraction $(c_P - c_V)/c_P$ of the energy supplied to the atmosphere does mechanical work against its pressure, while its internal energy is increased by the remaining fraction c_V/c_P. If the atmosphere were not permitted to move the two specific heats would be the same, and all of the energy injected by the explosion would appear as internal energy of the air. Equating W to the portion of the explosive energy which does mechanical work gives

$$\begin{aligned} Y &= \left(\frac{c_P}{c_P - c_V}\right) 2\pi R^2 \Delta R P_0 \\ &= \left(\frac{\gamma}{\gamma - 1}\right) 2\pi R^2 \Delta R P_0. \end{aligned} \qquad (A.2.2)$$

For nuclear explosions this is a conveniently measured displacement. If $Y = 4.18 \times 10^{20}$ erg (equivalent to 10 kilotons of standard conventional explosive) and $R = 10$ km, $\Delta R = 19$ cm. However, $\Delta R \propto R(\Delta P/P_0)$, where ΔP is the overpressure at the point of observations. For small explosions, where observations are conducted at small R, the displacements may be too small to measure easily.

A number of approximations have been made. We have neglected the energy coupled into the ground, which is very small because the large jump in density and sound speed at the surface leads to a large acoustic mismatch. The kinetic energy of the moving air is also small. Its characteristic velocity $v \sim \Delta R/t \sim c_s(\Delta R/R)$, where t is the time required for the shock to arrive after the explosion and c_s is the sound speed of air. From this it is easy to see that the fraction of W which appears as kinetic energy is $\mathcal{O}(Y/P_0R^3) \ll 1$. There are more significant errors. In the inner parts of the shocked air the temperature is very high and γ differs from (is generally less than) its value in cool air. The calculation also ignores the energy radiated by the hot fireball. For nuclear explosions in air this leads to an underestimate of Y by a factor ~ 2, and for shocks in the interstellar medium, which pass through a long snowplow stage (regime IIb in **3.4**) in which they radiate strongly, this error may be even larger. Finally, if the result is to be applied to a shock which expands spherically, as is the case in most astronomical problems, the factor 2π must be replaced by 4π.

In principle, this method could be applied to interstellar shocks, if the displacement of suitable light objects could be measured. Because interstellar shocks move so slowly (on astronomical distance scales) it would probably be necessary to measure the displacement of a continuous emitting gas filament where the shock had passed through it, in comparison to a place where a dense cloud shielded it from the shock. It is unlikely to be feasible to observe the displacement of the filament as it occurs.

Index

⑩

The Addison-Wesley **Advanced Book Program** would like to offer you the opportunity to learn about our new physics titles in advance. To be placed on our mailing list and receive pre-publication notices and discounts, just **fill out this card completely** and return to us, postage paid. Thank you.

Name ______________________

Title ______________________

School/Company ______________________

Department ______________________

Street Address ______________________

City ____________ State ________ Zip ________

Telephone (________) ______________________

Where did you buy this book?

- ☐ Bookstore (non-campus)
- ☐ Mail Order
- ☐ School/Required for Class
- ☐ Other ______________________
- ☐ Campus Bookstore/Individual Study
- ☐ Toll Free # to Publisher
- ☐ Professional Meeting

What professional physics & science associations are you an *active* member of?

- ☐ AAPT (American Association of Physics Teachers)
- ☐ AIP (American Institute of Physics)
- ☐ APS (American Physical Society)
- ☐ Sigma Pi Sigma
- ☐ SPS (Society of Physics Students)
- ☐ Other ______________________

Check your areas of interest.

- 11 ☐ Quantum Mechanics
- 12 ☐ Particle/Astro Physics
- 13 ☐ Condensed Matter/Solid State Physics
- 14 ☐ Mathematical Physics
- 15 ☐ Nuclear Physics
- 16 ☐ Electron & Atomic Physics
- 17 ☐ Plasma Physics
- 18 ☐ Materials Physics
- 19 ☐ Biological Physics
- 20 ☐ High Polymer Physics
- 21 ☐ Chemical Physics
- 22 ☐ Fluid Dynamics
- 23 ☐ History of Physics
- 24 ☐ Statistical Physics
- 25 ☐ Other ______________________

Are you more interested in: ☐ theory ☐ experimentation?

Are you currently writing or planning to write a physics book?

☐ Yes ☐ No

Area: ______________________

(If Yes) **Are you interested in discussing your project with us?**

☐ Yes ☐ No

Ktz

MAILING LIST • MAILING LIST • MAILING LIST • MAILING LIST • MAILING LIST • MAILING LIST

cut along dotted line

No Postage
Necessary
if Mailed in the
United States

BUSINESS REPLY CARD
FIRST CLASS PERMIT NO. 922 MENLO PARK, CA 94025

Postage will be paid by Addressee:

Addison-Wesley
Publishing Company, Inc.®
Advanced Book Program

2725 Sand Hill Road
Menlo Park, California 94025-9919